TWO-YEAR COLLEGE MATHEMATICS READINGS

Edited by Warren Page*

CONSULTING PANEL

Donald J. Albers
Edwin F. Beckenbach
Robert Bumcrot
Leo Chosid
Lawrence A. Curnutt
Thomas M. Green

Vedula N. Murty
Norman Schaumberger
J. Arthur Seebach
Lawrence A. Sher
Edward J. Specht
Lynn A. Steen

With the reviewing assistance of Glenn D. Allinger, Crist D. Dixon, Kay W. Dundas, Russell Griemsman, Louise . Grinstein, Roland H. Lamberson, Simon J. Lawrence, Don O. Loftsgaarden, Ely H. Stern, Alan C. Tucker, zelle B. Waltcher, Milton E. Winger, Roger L. Woodriff.

This work was partially supported by the Alfred P. Sloan Foundation Grant-in-Aid No. B1979-9.

Published and distributed by
THE MATHEMATICAL ASSOCIATION OF AMERICA

ISBN 0-88385-435-X

Printed in the United States of America

Current printing (last digit):

10 9 8 7 6 5 4 3 2 1

Foreword

A striking feature of American education during the mid-portion of the present twentieth century has been the phenomenal growth of our two-year colleges.

Recognizing the burgeoning importance of two-year-college mathematics, in 1970 the Mathematical Association of America cooperated in the founding of THE TWO-YEAR COLLEGE MATHEMATICS JOURNAL, and in 1974 MAA became sole publisher of TYCMJ and adopted the JOURNAL as one of its official publications.

For some time it has seemed desirable for the Association to give additional recognition of the place of mathematics in the two-year colleges by the separate publication of a volume of relevant articles, and thus further to serve the needs and interests of its large and dynamic two-year-college component. Fortunately, ample material was on hand in the abundance of articles already deemed acceptable for publication in TYCMJ.

The formidable task of selecting, organizing, and unifying material for READINGS was undertaken by Professor Warren Page of New York City Technical College. A period of released time for some of his work on the project was made possible by a generous and greatly appreciated Alfred P. Sloan Foundation grant, and his task was somewhat alleviated by the assistance of members of the Editorial Board that he organized.

To Professor Page, to the Sloan Foundation, and to all who were involved with this project, the Association hereby expresses its sincere appreciation for their contributions toward bringing this significant book into being.

E. F. BECKENBACH, Chairman
Committee on Publications

Contents

1. ALGEBRA
1

The Harmonic Mean and the EPA, *Clifford H. Wagner* . . . 3
A Combinatoric Identity with Applications, *Herman Atkinson and Warren Page* 5
The Greatest-Integer Function: A Function with Many Uses, *S. Avital and R. Hansen* 10
A Visual Approach to Continued Fractions, *D'Ann Fuquay* . . 18
Negative Bases with Rational Exponents, *David Rudd* . . . 22
Mathematical Induction and Matrix Inversion, *Robert M. Tardiff* 24

2. GEOMETRY
27

Pick's Theorem for Simple Closed Polygons, *J. B. Harkin* . . 30
Angle Trisection Using Compass and Finite Straightedge, *Joe Dan Austin* 36
On Some Properties of the Quadrangle, *Albert Schild* . . . 40
Tchebysheff Polynomials and Primitive Pythagorean Triples, *Gerald Bergum and Ken Yocom* 48
Semiregular Polyhedra, *Charles H. Jepsen* 51
Angles in Graphs, *Frank Harary and David B. Leep* 62

3. NUMBER THEORY
75

Initial Primes, *Richard L. Francis* 78
Mersenne Primes and the Lucas-Lehmer Test, *W. G. Leavitt* 80
An Introduction to the Tarry-Escott Problem, *Harold L. Dorwart and Warren Page* 87
The Digital Root Function, *Thomas P. Dence* 96

4. CALCULUS
105

Does the Formula for Arc Length Measure Arc Length?, *Kenneth P. Goldberg* 107
The Formula for Arc Length Does Measure Arc Length, *Warren Page* 111

Rational Solutions of $x^y = y^x$, *Charles S. Weaver* 115
A Vectorial Approach to "The Chase," *Sidney H. L. Kung* . . 118
Discontinuity Amidst Differentiability, *Michael W. Ecker* . . . 121

5. PROBABILITY & STATISTICS   133

Continuous Dice, *Lawrence J. Pozsgay* 136
Gamblers' Ruin, *Homer T. Hayslett, Jr.* 141
A Graph-Theoretic Approximation of e, *Bruce Golden and
    Daniel Casco* 147
The Orthogonal Least Squares Line, *David Farnsworth and
    Steven Suddaby* 149
A Statistical Experiment, *Neville Spencer* 156
Buffon, the Computer and the Petersburg Paradox, *David
    C. Tolman and James H. Foster* 161

6. CALCULATORS & COMPUTERS   171

Direct Reduction Loans, *John T. Varner III* 174
Calculators and Polynomial Evaluation, *J. F. Weaver* 176
Integrating Mathematics and Computer Science: A Case for
    LISP, *Michael E. Burke* 188
Simulating Sampling from Normal Populations, *Catherine
    Lilly* 194
Extended Precision Arithmetic–A Numeric Approach,
    Gerard Kiernan 198

7. MATHEMATICS EDUCATION   203

A Logical Approach to Teaching Formal Logic, *G. Vervoort* . 206
The Search for Sums, *S. Avital and Edward Theil* 209
An Elementary Proof for a Famous Counting Problem,
    James O. Friel 218
Problem Solving Versus Answer Finding, *John Staib* . . . 221
Using Monte Carlo Methods to Solve Probability Problems,
    Carolyn Shevokas Funk 228
Introducing the Computer in Liberal Arts Mathematics, *S. P.
    Gordon, R. W. Meyer, and A. Shindhelm* 233
Mathematics Placement Test Construction, *F. Pedersen* . . 239

8. RECREATIONAL MATHEMATICS   251

Aspects of Group Theory in Residue Designs, *Thomas E.
    Moore* 254

De Bruijn's Packing Problem, *Richard Johnsonbaugh* . . . 258
A General Approach to $p.q$ r-cycles, *Warren Page* 263
Coin Weighings Revisited, *Aaron L. Buchman and Frank S.*
    Hawthorne 275
How to Tree Instant Insanity®, *Jean J. Pedersen* 282
The Game Beware, *F. Pedersen and K. Pedersen* 291

1. ALGEBRA

> *"Arithmetical symbols are written diagrams and geometrical figures are graphic formulas."*
>
> —D. Hilbert, Mathematical Problems,
> Bull. Amer. Math. Soc. 8 (1902), 443

> *"That flower of modern mathematical thought— the notion of a function."*
>
> —Thomas J. McCormack, On the Nature of
> Scientific Law and Scientific Explanation,
> Monist 10 (1899–1900), 555

> *"We may always depend upon it that algebra, which cannot be translated into good English and sound common sense, is bad algebra."*
>
> —W. K. Clifford, Common Sense in the Exact Sciences
> (London, 1885), Chapter 1, Section 7

Algebra* may be described as generalized arithmetic since it is the study of arithmetic in which letters represent numbers. The arithmetic facts $3 + 3 = 2 \times 3$ and $7 + 7 = 2 \times 7$, for example, are particular cases of the algebraic statement that $x + x = 2x$ for every number x.

The use of letters to stand for numbers represents a major step toward abstraction and the solution of more complicated problems. This is illustrated by the first two articles of the chapter. Clifford Wagner's note "The Harmonic Mean and the EPA" describes how a simple algebraic formula can be used to explain how the Environmental Protection Agency determines average fuel economy for automobiles. And, in "A Combinatoric Identity with Applications," Herman Atkinson and Warren Page use an algebraic result to solve counting problems in set theory and modern algebra.

---

*The word "Algebra" derives from the Arabic title of Alkarismi's 9th century work "Hisâb al-jabr w'al Mugâbalah" dealing with solving equations and simplifying algebraic expressions.

1

Once letters are allowed to represent numbers, it seems only natural to express algebraic notions more compactly through the use of symbols—as, for example, $[x]$ denoting the greatest-integer less than or equal to x, and $\langle a_0, a_1, a_2 \rangle$ representing the continued fraction

$$a_0 + \cfrac{1}{a_1 + (1/a_2)}.$$

While they help to enhance algebra's efficiency and reduce its dependence on words, symbolic expressions increase abstraction and make it easy to become perplexed when one's intuition is bypassed. Thus, it becomes increasingly desirable to be able to visualize what is happening as algebraic expressions become more complex. The value of a combined symbolic-visual approach to understanding is quickly brought home in the next two articles. S. Avital and R. Hansen effectively use coordinate geometry to illustrate their survey "The Greatest-Integer Function: A Function with Many Uses." D'Ann Fuquay takes another perceptual route. In "A Visual Approach to Continued Fractions," she presents a pictorial interpretation of continued fractions—an interpretation that can be used to complement students' understanding of the algebraic nature of the division algorithm.

As does every system of study, algebra has its own internal constraints, and these limitations invite attempts to go beyond the confines of the study. For example, some laws of algebra are defined only for certain sets of numbers, and so it is tempting to seek less restricted sets of numbers for which these laws apply. The usual laws of exponents hold for negative bases with integral exponents. However, there is no way to define $a^{1/2}$ $(a < 0)$ as a real number and still satisfy $(a^u)^v = a^{uv}$ for all rational exponents, since this yields $0 < (a^{1/2})^2 = a^1 = a < 0$. Thus, a^u can be defined only for certain rational exponents. David Rudd's note "Negative Bases with Rational Exponents" describes the set of rational exponents that can be used to define a^u so that the usual laws of exponents apply.

Algebra, originally the generalized arithmetic of the ordinary number system, eventually evolved into the arithmetic of other mathematical systems in which numbers were replaced by objects. Matrix algebra, for example, encompasses number algebra (real numbers are 1×1 matrices), and so it is natural to expect that matrices will enjoy many, but not all (matrix multiplication is noncommutative) properties analogous to those of real numbers. Thus, "A square matrix is invertible if and only if its determinant is nonzero" may be viewed as a generalization of "A real number has an inverse if and only if it is nonzero." In "Mathematical Induction and Matrix Inversion," Robert Tardiff uses mathematical induction to establish the matrix analog of "The real number $1 - a$ $(a \geqslant 0)$ has an inverse $(1 - a)^{-1} > 0$ if $(1 - a)x > 0$ for some $x > 0$." This article is a fitting way to conclude the chapter since it compares and contrasts the evolutionary interconnections between the algebra of numbers and the algebra of matrices.

The Harmonic Mean and the EPA

Clifford H. Wagner, *Pennsylvania State University, Middletown, PA*

Trying to understand the government mileage ratings for new cars can be both enlightening and humbling. For example, consider the Environmental Protection Agency's ratings for the 1978 Volkswagen Rabbit with diesel engine: $c = 40$ miles per gallon in city driving, $h = 53$ mpg on the highway, and $a = 45$ mpg as an overall average of city and highway driving. How did they get the value 45 as an average of 40 and 53?

The simple mean,

$$\overline{X} = (40 + 53)/2 = 46.5 \text{ mpg},$$

is not appropriate because the EPA has determined that 55% of all driving is in the city, with 45% on the highway.

So, why not use the weighted mean,

$$\overline{X}_w = (.55)(40) + (.45)(53) = 45.85 \text{ mpg?}$$

This weighted mean would be correct only if 55% of all gasoline were *consumed* in city driving, with 45% *consumed* on the highway. For example, if that were the case, then for every 100 gallons of gasoline consumed, the Rabbit would travel $(55)(40) = 2200$ miles in the city and $(45)(53) = 2385$ miles on the highway, for a combined mileage of $(2200 + 2385)/100 = 45.85$ mpg.

Is it possible that the EPA uses an educated guess to obtain the overall average of c and h? A nice theorem of G. Pólya suggests that, in a certain sense, the best guess would be the harmonic mean

$$H = \frac{2ch}{c + h} = (.50/c + .50/h)^{-1} = 45.59 \text{ mpg}.$$

Pólya's theorem (see [1] or [3] for an elementary proof) asserts that: "*The approximation that yields the minimum for the greatest possible absolute value of the relative error, committed in approximating an unknown quantity contained between two known positive bounds, is the harmonic mean of these bounds.*"

Actually, if city and highway driving were weighted equally, the harmonic mean would yield the appropriate average. To justify this claim, let us consider the rate of consumption which is the inverse of mileage. The Rabbit consumes $c^{-1} = 1/40$ gallons per mile in city driving, and $h^{-1} = 1/53$ gallons per mile on the highway. Using equal weights for city and highway driving would mean that approximately half of each mile driven is in the city, where the car consumes $.50/c$ gallons, and half of each mile driven is on the highway, where the car consumes $.50/h$ gallons. Thus the overall rate of consumption would be $.50/c + .50/h$ gallons per mile. By taking inverses, we find that the corresponding overall mileage would be given by the harmonic mean.

Now let us answer our initial question. Since the proportions of city and highway driving are not equal, the appropriate average is the weighted harmonic mean. The weighted mean rate of *consumption* is $.55/c + .45/h$ gallons per mile, and the corresponding weighted harmonic mean for *mileage* is

$$H_w = (.55/c + .45/h)^{-1} = 44.96 \doteq 45 \text{ mpg.}$$

A similar application of the weighted harmonic mean actually appears in public law. The Motor Vehicle Information and Cost Savings Act [2], by which Congress has set yearly average fuel economy requirements for each automobile manufacturer, specifies that: *For each manufacturer, the mileage ratings of all different models are to be averaged using the weighted harmonic mean, with each model weighted according to its proportion of the manufacturer's total production.* The legal recipe for this calculation appears below.

PUBLIC LAW 94-163

"DETERMINATION OF AVERAGE FUEL ECONOMY

15 USC 2003.   "SEC. 503. (a) (1) Average fuel economy for purposes of section 502 (a) and (c) shall be calculated by the EPA Administrator by dividing—
"(A) the total number of passenger automobiles manufactured in a given model year by a manufacturer, by
"(B) a sum of terms, each term of which is a fraction created by dividing—
"(i) the number of passenger automobiles of a given model type manufactured by such manufacturer in such model year, by
"(ii) the fuel economy measured for such model type.

REFERENCES

1. N. D. Kazarinoff, Analytic Inequalities, Holt, Rinehart, and Winston, New York, 1961, 39.
2. Motor Vehicle Information and Cost Savings Act, U.S. Code, title 15, section 2003 (1975).
3. G. Pólya, On the harmonic mean of two numbers, Amer. Math. Monthly, 57 (1950) 26–28; reprinted in Selected Papers on Calculus, ed. Tom M. Apostol *et al.*, Math. Assoc. Amer., 1969, 254–256.

A Combinatoric Identity with Applications

Herman Atkinson, *Seward, NE* and Warren Page, *New York City Technical College, Brooklyn, NY*

The following identity has many interesting and useful applications: *If p is a prime number, then*

$$\left[\frac{\sum_{j=0}^{k} a_j p^j}{\sum_{j=0}^{k} b_j p^j} \right] \equiv \binom{a_0}{b_0}\binom{a_1}{b_1} \cdots \binom{a_k}{b_k} \ (\text{mod } p) \qquad (*)$$

for each set of nonnegative integers $\{a_j, b_j < p : 0 \leqslant j \leqslant k\}$.

One surprisingly simple application of $(*)$ concerns the order (i.e., cardinality) $°(S)$ of a set S.

If $p^\alpha | °(S)$, there exists $n \equiv 1$ (mod p) subsets of S that have order p^α and contain a fixed element of S.

$$(**)$$

And using $(**)$, we obtain a distinctly new proof of the well-known Sylow Theorem (see [1], [2]):

If p is prime and $p^\alpha | °(G)$, the group G has $n \equiv 1$ (mod p) subgroups of order p^α.

$$(***)$$

Proof of $(*)$[†] Begin with the standard identity (proved by induction)

$$\binom{n+1}{r} = \binom{n}{r} + \binom{n}{r-1}, \qquad (1)$$

and use induction on k to establish that

$$\binom{n+k}{r} = \sum_{i=0}^{k} \binom{k}{i}\binom{n}{r-i}. \qquad (2)$$

[†]The authors are grateful to Crist Dixon for his considerable assistance with the proof of $(*)$.

Next, observe that

$$\binom{p^k}{i} \equiv 0 \pmod{p} \quad \text{for } i = 1, 2, \ldots, p^k - 1 \tag{3}$$

follows from

$$\binom{p^k}{i} = \frac{p^k(p^k - 1)!}{i(i-1)!\{p^k - 1) - (i-1)\}!} = \frac{p^k}{i}\binom{p^k - 1}{i - 1}.$$

We are now ready to prove, by induction on r, that

$$\binom{rp^k + \alpha_k}{sp^k + \beta_k} \equiv \binom{r}{s}\binom{\alpha_k}{\beta_k} \quad \pmod{p} \tag{$*_k$}$$

for nonnegative integers r, s, α_k, β_k with $\alpha_k < p^k$. Evidently $(*_k)$ holds for $r = 0$ and $s = 0$. For $r = 0$ and $s \geqslant 1$, we have

$$\alpha_k < p^k \leqslant sp^k + \beta_k$$

and so

$$\binom{\alpha_k}{sp^k + \beta_k} = 0 = \binom{0}{s}.$$

Thus, $(*_k)$ holds for $r = 0$ and any $s \geqslant 0$. Assume that $(*_k)$ holds for any fixed r and all $s \geqslant 0$. Then

$$\begin{bmatrix} (r+1)p^k + \alpha_k \\ sp^k + \beta_k \end{bmatrix} = \begin{bmatrix} (rp^k + \alpha_k) + p^k \\ sp^k + \beta_k \end{bmatrix}$$

$$= \sum_{i=0}^{p^k} \binom{p^k}{i}\binom{rp^k + \alpha_k}{sp^k + \beta_k - i} \qquad [\text{via (2)}]$$

$$\equiv \left\{\binom{rp^k + \alpha_k}{sp^k + \beta_k} + \begin{bmatrix} rp^k + \alpha_k \\ (s-1)p^k + \beta_k \end{bmatrix}\right\} \pmod{p} \quad [\text{via (3)}]$$

$$\equiv \left\{\binom{r}{s}\binom{\alpha_k}{\beta_k} + \binom{r}{s-1}\binom{\alpha_k}{\beta_k}\right\} \pmod{p} \qquad [\text{by hypothesis}]$$

$$\equiv \binom{r+1}{s}\binom{\alpha_k}{\beta_k} \pmod{p}. \qquad [\text{via (1)}]$$

This establishes $(*_k)$ for each fixed integer $k \geqslant 0$.

It is a simple matter now to prove $(*)$ by induction on k. Clearly $(*)$ holds for $k = 0$. To see that $(*)$ holds for $k = 1$, invoke $(*_1)$ with $\alpha_1 = a_0$ and $\beta_1 = b_0$. Similarly, $(*)$ holds for $k = 2$ since $(*_2)$ holds with $\alpha_2 = a_1 p + b_0$ and $\beta_2 = b_1 p + b_0$, and $(*)$ has previously been verified for

$k = 1$. Assume that $(*)$ holds for $k = 0, 1, \ldots, K - 1$. Taking

$$\alpha_K = \sum_{j=0}^{K-1} a_j p^j \quad \text{and} \quad \beta_K = \sum_{j=0}^{K-1} b_j p^j,$$

and noting that $\alpha_K \leqslant \sum_{j=0}^{K-1}(p-1)p^j = p^K - 1$, we obtain

$$\begin{pmatrix} \sum_{j=0}^{K} a_j p^i \\ \sum_{j=0}^{K} b_j p^j \end{pmatrix} = \begin{pmatrix} a_K p^K + \alpha_K \\ b_K p^K + \beta_K \end{pmatrix} \equiv \begin{pmatrix} a_K \\ b_K \end{pmatrix}\begin{pmatrix} \alpha_K \\ \beta_K \end{pmatrix} (\bmod\ p) \quad [\text{via } (*_K)]$$

$$\equiv \begin{pmatrix} a_K \\ b_K \end{pmatrix} \prod_{j=0}^{K-1} \begin{pmatrix} a_j \\ b_j \end{pmatrix} (\bmod\ p) \quad [\text{by inductive hypothesis}]$$

$$\equiv \begin{pmatrix} a_0 \\ b_0 \end{pmatrix}\begin{pmatrix} a_1 \\ b_1 \end{pmatrix} \cdots \begin{pmatrix} a_K \\ b_K \end{pmatrix} (\bmod\ p).$$

This completes the proof of $(*)$.

Proof of $(**)$. Assume that $°(S) = m \cdot p^\alpha$, and let E be the family of subsets of S that have p^α elements and contain a fixed $x \in S$. Then

$$°(E) = \begin{pmatrix} m \cdot p^\alpha - 1 \\ p^\alpha - 1 \end{pmatrix}$$

$$= \begin{bmatrix} (p-1) + (p-1)p + \cdots + (p-1)p^{\alpha-1} + (m-1)p^\alpha \\ (p-1) + (p-1)p + \cdots + (p-1)p^{\alpha-1} + 0 \cdot p^\alpha \end{bmatrix}$$

$$\equiv 1 (\bmod\ p).$$

Proof of $(***)$. Let $°(G) = m \cdot p^\alpha$, and let E be the family of subsets of G that have p^α elements (and contain the identity element $e \in G$). Define an equivalence relation on E by defining $M, M' \in E$ to be equivalent if $M' = Mg^{-1}$ for some $g \in M$. Now consider any equivalence class $\overline{M} = \{Mg^{-1} \in E : g \in M\}$, and let $H = \{h \in G : Mh = M\}$. It is easily verified that H is a subgroup of G. Furthermore, $h = eh \in Mh = M$ for each $h \in H$. Therefore, $H \subset M$ and $1 \leqslant °(H) \leqslant p^\alpha$. There are two cases to consider:

$$°(H) < p^\alpha \quad \text{or} \quad °(H) = p^\alpha.$$

Suppose $°(H) < p^\alpha$. Then $H \subsetneq M$ and there is a nonidentity element $g_2 \in M - H$. Clearly $g_2 H \neq H$, and $g_2 H \subset MH = M$. If $H \cup g_2 H \neq M$, there is another element $g_3 \in M - (H \cup g_2 H)$ such that $g_3 H \neq H$ and $g_3 H \subset M$. Proceeding in this manner, we obtain

$$M = g_1 H \cup g_2 H \cup \cdots \cup g_k H \tag{3}$$

where

$$g_1 = e \quad \text{and} \quad g_j \in M - \bigcup_{i=1}^{j-1} g_i H \quad (j = 2, 3, \ldots, k).$$

If $g \in M$, then $g = g_i h$ for one of the above g_i's and some $h \in H$. Therefore $Mg^{-1} = Mh^{-1}g_i^{-1} = Mg_i^{-1}$, and we see that $\overline{M} = \{ Mg_i^{-1} : 1 \leqslant i \leqslant k \}$. These members of $\overline{M}$ are also distinct since $Mg_i^{-1} = Mg_j^{-1} (j > i)$ means that $g_j^{-1}g_i \in H$, and $g_j^{-1} \in Hg_i^{-1}$ yields the contradiction that $g_j \in g_i H^{-1} = g_i H$. Hence $°(\overline{M}) = k$. But $p^\alpha = k \cdot °(H)$ from (3). Since $°(H) < p^\alpha$, we have $p \mid k$ and $°(\overline{M}) \equiv 0 \pmod{p}$. This, however, cannot be true for every equivalence class in E since we would then have $°(E) \equiv 0 \pmod{p}$, which contradicts $°(E) \equiv 1 \pmod{p}$ established by $(\ast\ast)$.

The preceding remarks demonstrate that $°(H)$ necessarily equals p^α for at least one member $M \in E$. This means that $H = M$ and $\overline{M} = \{ Hg^{-1} : g \in H \} = \{H\}$ has $°(\overline{M}) = 1$. Therefore, the number of classes for which this is the case must be $n \equiv 1 \pmod{p}$, since we could not have $°(E) \equiv 1 \pmod{p}$ if this were not so.

Example. Let the vertices of a square be numbered as shown below.

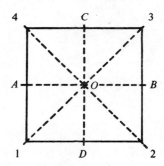

Let e, r, s, t denote counterclockwise rotations of the square about O of 0, 90, 180, and 270 degrees, respectively. These rotations give rise respectively to $e = (1)$, $r = (1234)$, $s = (13)(24)$, and $t = (1432)$. The rotations through 180 degrees about the bisectors AOB and COD give rise respectively to $a = (14)(23)$ and $b = (12)(34)$, while the rotations through 180 degrees about the diagonals 204 and 103 give rise to $c = (13)$ and $d = (24)$. The symmetries of the square

$$e = (1) \qquad r = (1234) \qquad s = (13)(24) \quad t = (1432)$$
$$a = (14)(23) \quad b = (12)(34) \quad c = (13) \qquad d = (24)$$

determine the well-known octic or dihedral group D_8 whose composition table is

D_8	e	r	s	t	a	b	c	d
e	e	r	s	t	a	b	c	d
r	r	s	t	e	c	d	b	a
s	s	t	e	r	b	a	d	c
t	t	e	r	s	d	c	a	b
a	a	d	b	c	e	s	t	r
b	b	c	a	d	s	e	r	t
c	c	a	d	b	r	t	e	s
d	d	b	c	a	t	r	s	e

Here $°(D_8) = 8 = m \cdot p^\alpha$ for $m = p = \alpha = 2$. Accordingly, E consists of the following $\binom{7}{3} = 35$ subsets of D_8 that have $p^\alpha = 4$ elements and contain e:

$erst$	$ertc$	$ercd$	$esad$	$etbc$
$ersa$	$ertd$	$esta$	$esbc$	$etbd$
$ersb$	$erab$	$estb$	$esbd$	$etcd$
$ersc$	$erac$	$estc$	$escd$	$eabc$
$ersd$	$erad$	$estd$	$etab$	$eabd$
$erta$	$erbc$	$esab$	$etac$	$eacd$
$ertb$	$erbd$	$esac$	$etad$	$ebcd$

There are thirteen equivalence classes $\overline{M} = \{ Mg^{-1} : g \in M \}$ into which E is partitioned by the equivalence relation on E:

$$\{ \overline{erst} \} = \{ erst \} \qquad \{ \overline{erac} \} = \{ erac, tecb \}$$

$$\{ \overline{esab} \} = \{ esab \} \qquad \{ \overline{erad} \} = \{ erad, teca \}$$

$$\{ \overline{escd} \} = \{ escd \} \qquad \{ \overline{ercb} \} = \{ ercb, tedb \}$$

$$\{ \overline{erbd} \} = \{ erbd, teda \}$$

$$\{ \overline{ersa} \} = \{ ersa, terc, steb, acbe \}$$

$$\{ \overline{ersb} \} = \{ ersb, terd, stea, bdae \}$$

$$\{ \overline{ersc} \} = \{ ersc, terb, sted, cbde \}$$

$$\{ \overline{ersd} \} = \{ ersd, tera, stec, dace \}$$

$$\{ \overline{erab} \} = \{ erab, tecd, aces, bsde \}$$

$$\{ \overline{ercd} \} = \{ ercd, teba, cbes, dase \}$$

It is now clear that there are three subgroups H having $°(H) = p^\alpha = 4$, these being $\{ e, r, s, t \}$, $\{ e, s, a, b \}$, and $\{ e, s, c, d \}$.

REFERENCES

1. W. Ledermann, Introduction to the Theory of Finite Groups, Interscience, New York, 1961, pp. 126–135.
2. Louis Shapiro, Introduction to Abstract Algebra, McGraw-Hill, Manchester, Missouri, 1975, pp. 172–175.

The Greatest-Integer Function: A Function with Many Uses

S. Avital, *I.I.T. Technion, Haifa, Israel,* and R. Hansen, *Montana State University, Bozeman, MT*

Greatest-integer functions have many interesting properties and applications. They can be effectively used for pedagogical purposes and for illustrating "real world" applications of mathematics.

Let **I** denote the set of integers. The greatest-integer function is defined on the set of real numbers as $d(x) = [x]$, where

$$[x] = \begin{cases} x, & x \in \mathbf{I}, \\ \text{the greatest integer less than } x, & x \notin \mathbf{I}. \end{cases}$$

Thus, for example, we have $[3] = 3 = [\pi]$ and $[-2] = -2 = [-1.01]$. The function $d(x) = [x]$ also describes what is called "truncation" in computer work. Here the machine "cuts off" the fractional part of the number, thus leaving only the integral part.

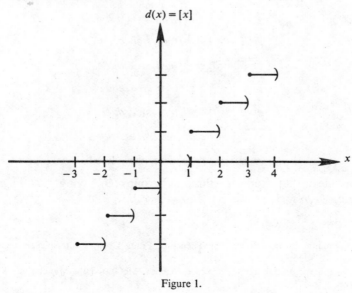

$d(x) = [x]$

Figure 1.

10

Undoubtedly the most distinguishing visual feature of the greatest-integer function is the jumps in its graph (Figure 1).

The greatest-integer function appears frequently in everyday life. For example, the change from \$1 that you receive when you buy one item selling at 3 for \$1 is $[100 - 100/3] = [66\frac{2}{3}] = 66$¢. Instead of using $[x]$ to "round down" (the change received), one may use $-[-x]$ to "round up" (the price paid). Thus, you pay $-[-100/3] = -[-33\frac{1}{3}] = -(-34) = 34$¢ for one item in this example.

In general, the function $U(x) = -[-x]$ "rounds up" by assigning the least-integer greater than or equal to the real number x (see Figure 2). This function $U(x)$ is used at the post office where one is charged the higher rate "for an additional unit (ounce, gram, pound, kilogram) or *any part thereof*."

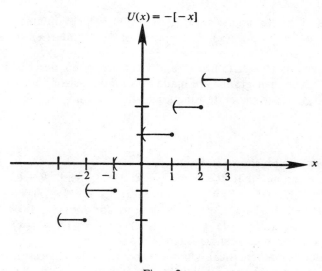

Figure 2.

Using $d(x) = [x]$, we defined $U(x) = -d(-x)$. In similar fashion, other greatest-integer type functions can be generated. We begin by considering

$$\Delta(x) = [x] + [-x] \quad \text{and} \quad s(x) = [x] - [-x],$$

so labeled since $\Delta(x) = d(x) - U(x)$ and $s(x) = d(x) + U(x)$. It is readily observed that $\Delta(n) = 0$ and $s(n) = 2n = 2[n]$ for every integer n. If $0 < \epsilon < 1$, then

$$\Delta(n + \epsilon) = [n + \epsilon] + [-(n + \epsilon)] = n + \{-(n + 1)\} = -1$$

and

$$s(n + \epsilon) = [n + \epsilon] - [-(n + \epsilon)] = n - \{-(n + 1)\}$$
$$= 2n + 1 = 2[n + \epsilon] + 1.$$

For $-1 < \epsilon < 0$, we have

$$\Delta(n + \epsilon) = [n + \epsilon] + [-(n + \epsilon)] = (n - 1) + (-n) = -1$$

and

$$s(n + \epsilon) = s((n - 1) + (1 + \epsilon))$$
$$= [(n - 1) + (1 + \epsilon)] - [-\{(n - 1) + (1 + \epsilon)\}]$$
$$= (n - 1) - \{-(n)\} = 2n - 1 = 2[n + \epsilon] + 1.$$

The foregoing remarks establish that

$$\Delta(x) = \left\{ \begin{array}{ll} 0, & x \text{ integer,} \\ -1, & x \text{ noninteger,} \end{array} \right\} \quad \text{and} \quad s(x) = \left\{ \begin{array}{ll} 2[x], & x \text{ integer,} \\ 2[x] + 1, & x \text{ noninteger.} \end{array} \right\}$$

This gives a better analysis of $\Delta(x) + s(x) = 2[x]$, obtained from the definitions of $\Delta(x)$ and $s(x)$. The graphs of $\Delta(x)$ and $s(x)$ are shown in Figure 3.

The functions $d(x), U(x), \Delta(x)$ and $s(x)$ are often described as "step functions" since their graphs are made up of (equal-sized) horizontal steps. It is also possible to generate functions which are not step functions. For instance,

$$f(x) = x - [x]$$

has the "saw" graph shown in Figure 4.

Note, for example, that $f(2.3) = 2.3 - [2.3] = .3$. And, $f(-2.3) = (-2.3) - [-2.3] = (-2.3) - (-3) = .7$ is the same as $f(-3 + .7) = (-3 + .7) - [-3 + .7] = (-3 + .7) - (-3) = .7$. In general, for any x and $0 < \epsilon < 1$, we have $f(x + \epsilon) = (x + \epsilon) - [x + \epsilon] = (x + \epsilon) - x = \epsilon$. Thus, $f(x)$ assigns to each real number x its positive fractional part.

When doing mathematical computations, it is often necessary to round off to the nearest integer. Obviously, the function $r(x)$ that rounds x to the nearest integer does not always agree with $d(x) = [x]$ or with $U(x) = -[-x]$. For instance, $r(2.6) = 3 = U(2.6)$ does not agree with $d(2.6) = 2$, and $r(2.4) = 2 = d(2.4)$ does not agree with $U(2.4) = 3$. A number which is equally distant from two consecutive integers (therefore, $f(x) = .5$) is usually approximated by the greater integer. Accordingly, $5.5 \approx 6$ and $-3.5 \approx -3$. It is a simple matter to express $r(x)$ in terms of greatest-integer functions. To be sure, $r(x) = [x]$ if $x - [x] < 1/2$, and $r(x) = [x] + 1$ if $x - [x] \geqslant 1/2$. But this is exactly how $[x + \frac{1}{2}]$ acts on real values x. Therefore, $r(x) = [x + \frac{1}{2}]$. The graph of $r(x)$ is the step function in Figure 5.

The preceding discussion serves as good motivation for students to explore some interesting greatest-integer function properties. Although $[x + \frac{1}{2}] \neq [x] + \frac{1}{2}$ in general (verify this graphically and algebraically),

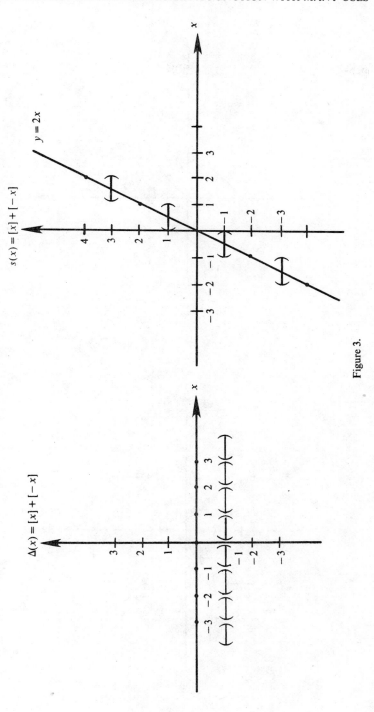

Figure 3.

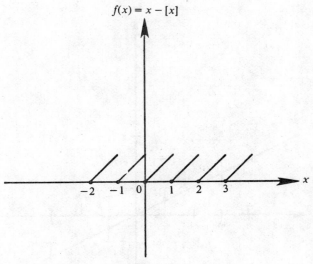

Figure 4.

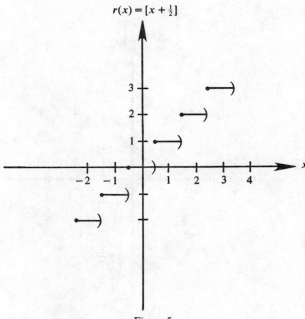

Figure 5.

students can (and should!) prove the following:

(a) $[x + n] = [x] + n$ for all integers n.
(b) $[x] + [y] \leqslant [x + y] \leqslant [x] + [y] + 1$.
(c) $[nx] \geqslant n[x]$ for all natural numbers n.
(d) $[[x]/n] = [x/n]$ for all natural numbers n.

Why do (c), (d) fail for arbitrary integers n? Graphs should be drawn whenever possible; some of these will be surprising. For example, see Figure 6.

Greatest-integer functions may be compared with other well-known functions—as, for example, $a(x) = |x|$. What, in contrast to Figure 6, does the graph $|x| + |y| = 1$ look like?

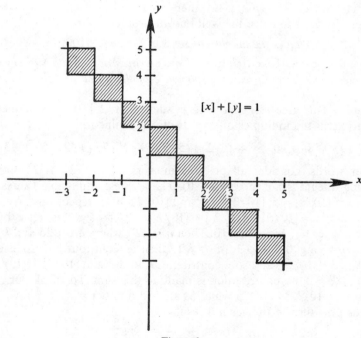

Figure 6.

Other Applications of the Greatest-Integer Function

A basic algorithm defined within the set of integers is the *division algorithm*: for integers a and b ($b > 0$), there exist unique integers q and r ($0 \leqslant r < b$) such that

$$a = bq + r.$$

This implies that

$$\frac{a}{b} = q + \frac{r}{b}.$$

One can immediately see that

$$\left[\frac{a}{b}\right] = \left[q + \frac{r}{b}\right] = q$$

since $q \in I$ and $0 \leqslant r/b < 1$. Hence we may write the division algorithm as

$$a = b\left[\frac{a}{b}\right] + r \quad (0 \leqslant r < b).$$

For example, $23 = 5[23/5] + 3$. Similarly, $-101 = 7[-101/7] + 4$.

Given integers $0 < p < q$, we may use the greatest-integer function to find the number of positive integers not exceeding q that are divisible by p. This requires that the numbers are $\{p, 2p, 3p, \ldots, sp\}$, where $sp \leqslant q < (s+1)p$. Since $s \leqslant q/p < s + 1$, there are $[q/p]$ such multiples of p. This result can be used to solve the well-known problem:

"*What is the number of zeros at the end of* $100!$?"

To solve the problem, we first ask: *If* $n!$ *is written as a product of powers of primes, what will be the greatest exponent* e *of the prime* p *such that* $p^e \mid n!$?

Example. The greatest exponent e such that $2^e \mid 10!$ is obtained by considering all the multiples of powers of 2 contained in

$$10! = 10 \cdot 9 \cdot 8 \cdot 7 \cdot 6 \cdot 5 \cdot 4 \cdot 3 \cdot 2 = (5 \cdot 2) \cdot 9 \cdot 2^3 \cdot 7 \cdot (3 \cdot 2) \cdot 5 \cdot 2^2 \cdot 3 \cdot 2.$$

In other words, consider all possible integers $q \cdot 2$, $r \cdot 2^2$, $s \cdot 2^3$ provided each such integer is not more than 10. The greatest admissible values of q, r, and s are $[10/2] = 5$, $[10/2^2] = 2$, and $[10/2^3] = 1$, respectively. We claim that, in fact, $e = [10/2] + [10/2^2] + [10/2^3] = 8$. To see this, let n be any positive integer not exceeding 10. Then $n = a2^k$, where a is odd and k is an integer satisfying $0 \leqslant k \leqslant 3$. If $k \geqslant 1$, then n contributes 1 to each of $[10/2^1], \ldots, [10/2^k]$, and so n contributes k to the sum $[10/2] + [10/2^2] + [10/2^3]$. If $k = 0$, no contribution is made to this sum. To check our result, we note that $10!/2^8 = 14175$ which is an odd number.

Let us now turn to the general result.

Theorem. *The greatest exponent* e *of the prime* $p \leqslant n$ *such that* $p^3 \mid n!$ *is*

$$e = \left[\frac{n}{p}\right] + \left[\frac{n}{p^2}\right] + \cdots + \left[\frac{n}{p^2}\right],$$

where $p^r \leqslant n < p^{r+1}$.

Proof. By the following observations:

(i) The factor p^e in $n!$ is obtained from various multiples of the form $q \cdot p^i$ which are not greater than n.

(ii) For each $1 \leqslant k \leqslant r$, the multiples of p^k which divide $n!$ are those which do not exceed n. These multiples are $p^k, 2p^k, \ldots,$ $[n/p^k]p^k$.

(iii) Each multiple $a = q \cdot p^k$ ($p \nmid q$) contributes an addend k to the exponent e.

(iv) Each such multiple $a = q \cdot p^k$ is counted precisely once by each addend $[n/p], [n/p^2], \ldots, [n/p^k]$, and is counted zero times by each addend of the form $[n/p^{k+1}], \ldots, [n/p^r]$.

(v) The exponent e is therefore equal to $[n/p] + [n/p^2] + \cdots + [n/p^r]$.

We can now solve the problem: "*How many zeros are there at the end of* 100!?" *Each zero will be obtained from a product of 2 times 5. As there are plenty of factors 2, we have to count only the number of factors 5. This number is* $[100/5] + [100/5^2] = 24$.

In similar fashion, one can find the highest power e of 6 such that $6^e \mid 100!$. Here $e = [100/3] + [100/3^2] + [100/3^3] + [100/3^4] = 33 + 11 + 3 + 1 = 48$. (Note: We again use the fact that there are abundantly many factors 2.)

Most number-theory texts provide a good deal of information about greatest-integer functions. Several such sources are listed below.

REFERENCES

1. C. T. Long, Elementary Introduction to Number Theory, D. C. Heath, Boston, Mass., 1965.
2. I. Niven and H. Zuckerman, An Introduction to the Theory of Numbers, Second Edition, John Wiley, New York, 1966.
3. J. Roberts, Elementary Number Theory, A Problem-Oriented Approach, MIT Press, Cambridge, Mass., 1977.
4. D. O. Shklarsky, N. N. Chentzov and I. M. Yaglom, The U.S.S.R. Olympiad Problem Book, W. H. Freeman, San Francisco, 1962.
5. J. Uspensky and M. Heaslet, Elementary Number Theory, McGraw-Hill, New York, 1936.

A Visual Approach to Continued Fractions

D'Ann Fuquay, *Mercer University, Macon, GA*

Students readily understand how to use the Euclidean algorithm to express rational numbers as continued fractions. It is clear to them how

$$51 = 2(22) + 7 \qquad 22 = 3(7) + 1 \qquad 7 = 7(1)$$

leads to

$$\frac{51}{22} = 2 + \frac{7}{22} = 2 + \frac{1}{\frac{22}{7}} = 2 + \frac{1}{3 + \frac{1}{7}}.$$

In more general terms, a rational number d_0/d_1 with $(d_0, d_1) = 1$ can be expressed as

$$
\begin{aligned}
d_0 &= a_0 d_1 + d_2 & 0 &< d_2 < d_1 \\
d_1 &= a_1 d_2 + d_3 & 0 &< d_3 < d_2 \\
d_2 &= a_2 d_3 + d_4 & 0 &< d_4 < d_3 \\
&\cdots\cdots & &\cdots\cdots \\
&\cdots\cdots & &\cdots\cdots \\
d_{n-1} &= a_{n-1} d_n + d_{n+1} & 0 &< d_{n+1} < d_n \\
d_n &= a_n d_{n+1} &
\end{aligned}
$$

and then rewritten

$$
\frac{d_0}{d_1} = a_0 + \cfrac{1}{a_1 + \cfrac{d_3}{d_2}} = a_0 + \cfrac{1}{a_1 + \cfrac{1}{a_2 + \cfrac{d_4}{d_3}}}
$$

$$\cdots\ \cdots\ \cdots$$
$$\cdots\ \cdots\ \cdots$$

$$
= a_0 + \cfrac{1}{a_1 + \cfrac{1}{a_2 + \cfrac{}{\ddots \ + \cfrac{1}{a_n}}}}.
$$

The integers a_i are called the *partial quotients* of d_0/d_1 since they are the quotients in the division algorithm. We offer the following visual interpretation of continued fractions, and their partial quotients, which can be used to complement students' understanding of the algebraic nature of this scheme.

Finite continued fractions

Standard notation will be used—namely,

$$\langle a_0, a_1, a_2, \ldots, a_n \rangle = a_0 + \cfrac{1}{a_1 + \cfrac{1}{a_2 + \cfrac{\ddots}{+ \cfrac{1}{a_n}}}}.$$

Every finite continued fraction represents a unique rational number. The converse is almost true (see [3], for example): Every rational number has two unique representations $\langle a_0, a_1, \ldots, a_n \rangle$ and $\langle a_0, a_1, \ldots, a_{n-1} - 1, 1 \rangle$. Thus, $51/22 = \langle 2, 3, 7 \rangle$ and $51/22 = \langle 2, 3, 6, 1 \rangle$. Our intent is to visually interpret the representation $\langle a_0, a_1, \ldots, a_n \rangle$ of d_0/d_1, where $(d_0, d_1) = 1$.

On a line segment of length $d_0 + d_1$, label the points with coordinates 0, d_0, $d_0 + d_1$ as Q_0, B, and Q_1, respectively. Thus $Q_0 B$ has length $|Q_0 B| = d_0$ and $Q_1 B$ has length $|Q_1 B| = d_1$.

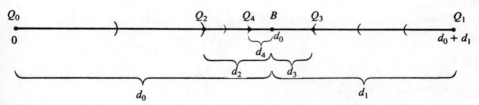

Figure 1.

Let a_0 denote the number of adjacent segments of length d_1 that can be marked off on $Q_0 B$ beginning at Q_0. If $d_0 > d_1$, then $a_0 \geqslant 1$. Denote the last point marked on $Q_0 B$ as Q_2, and let $d_2 = |Q_2 B|$.

If $d_2 = 0$, then $d_0 = a_0 d_1$ and $d_0/d_1 = a_0 = \langle a_0 \rangle$.

If $d_2 \neq 0$, we have

$$d_0 = a_0 d_1 + d_2 \qquad (0 < d_2 < d_1). \qquad (1)$$

(For the case in which $d_0 < d_1$, we have $a_0 = 0$. Simply relabel Q_0 and d_0 as Q_2 and d_2, and proceed.) Now place the compass at point Q_1, and mark off as many adjacent segments of length d_2 as possible on $Q_1 B$. Since $d_2 < d_1$, this number of marked off segments is $a_1 \geqslant 1$. Denote the last mark made as point Q_3, and let $d_3 = |Q_3B|$.

If $d_3 = 0$, then $d_1 = a_1 d_2$ and $d_1/d_2 = a_1$. Using (1), this gives $d_0/d_1 = a_0 + (1/a_1) = \langle a_0, a_1 \rangle$.

If $d_3 \neq 0$, we have

$$d_1 = a_1 d_2 + d_3 \qquad (0 < d_3 < d_2). \tag{2}$$

Place the compass at Q_2, and mark off as many adjacent segments of length d_3 as possible on Q_2B. Since $d_3 < d_2$, the number of marked off segments is $a_2 \geqslant 1$. Denote the last mark made as point Q_4, and let $d_4 = |Q_4B|$.

If $d_4 = 0$, then $d_2 = a_2 d_3$. Invoking (1) and (2), we obtain

$$\frac{d_0}{d_1} = a_0 + \cfrac{1}{a_1 + \cfrac{1}{a_2}} = \langle a_0, a_1, a_2 \rangle.$$

If $d_4 \neq 0$, we have

$$d_2 = a_2 d_3 + d_4 \qquad (0 < d_4 < d_3). \tag{3}$$

In general, the number of adjacent segments of length $d_k \neq 0$ which can be marked off on $Q_{k-1}B$ is $a_k = [d_k/d_{k+1}]$, where $[y]$ denotes the greatest-integer less than or equal to y. The point Q_{k+1} is the endpoint nearest B of the last marked off segment, and $d_{k+1} = |Q_{k+1}B| < d_k$. The construction can be continued in this manner until we arrive at a_n corresponding to $d_{n+2} = 0$.

Example.   Consider the case where $(d_0/d_1) = 7/22$.

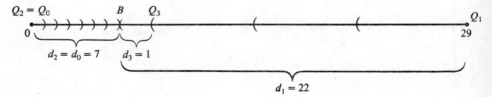

Figure 2.

Since $d_0 < d_1$, we have $a_0 = [d_0/d_1] = 0$. Accordingly, Q_0 and d_0 are relabeled Q_2 and d_2. Clearly, $a_1 = [d_1/d_2] = 3$ and $d_3 = |Q_3B| = 1$. Therefore $a_2 = [d_2/d_3] = 7$ and $d_4 = 0$. This yields $7/22 = \langle 0, 3, 7 \rangle$.

Infinite continued fractions

It is well known (for example, see [3]) that every infinite continued fraction represents an irrational number. Conversely, every irrational number x can be uniquely expressed as an infinite continued fraction $\langle a_0, a_1, a_2, \ldots \rangle$. The standard procedure is as follows: Let $\xi_0 = x$. Then define $a_0 = [\xi_0]$ and $\xi_1 = 1/(\xi_0 - a_0)$. Next, define $a_1 = [\xi_1]$ and $\xi_2 = 1/(\xi_1 - a_1)$. Proceed inductively to define

$$a_i = [\xi_i] \quad \text{and} \quad \xi_{i+1} = \frac{1}{\xi_i - a_i} \qquad (i = 0, 1, 2, \ldots).$$

The ξ_{i+1} are all irrational since the irrationality of ξ_i implies that of ξ_{i+1}. Furthermore, $\xi_{i+1} > 1$ and $a_{i+1} = [\xi_{i+1}] \geq 1$ for all $i \geq 0$. And, it can be shown (see [3], for example) that

$$x = \lim_{n \to \infty} \langle a_0, a_1, \ldots, a_n \rangle = \langle a_0, a_1, \ldots \rangle.$$

Our visual interpretation of continued fractions for rationals carries over intact to irrationals; the number of steps in this visual process (using $d_0 = x$ and $d_1 = 1$) simply becomes countably infinite. Clearly a_0, as defined by the visual construction, is the same a_0 defined in the continued fraction algorithm above. The continued fraction algorithm defines $\xi_1 = 1/(\xi_0 - a_0)$, and this is precisely d_1/d_2 in our visual process. Now

$$\xi_2 = \frac{1}{\xi_1 - a_1} = \frac{1}{\dfrac{d_1}{d_2} - a_1} = \frac{1}{d_1 - a_1 d_2} = \frac{d_2}{d_3}$$

(using (2) in the last equality). Similarly, (3) yields

$$\xi_3 = \frac{1}{\xi_2 - a_2} = \frac{d_3}{d_4}.$$

In general, $\xi_n = d_n/d_{n+1}$ follows inductively. Therefore each $a_n = [d_n/d_{n+1}]$ determined visually agrees with $a_n = [\xi_n]$ defined by the continuous fraction algorithm.

For further reading on continued fractions, see [1], [2], and [4].

REFERENCES

1. F. D'Ann Fuquay, The Sós Geometry of Continued Fractions, Doctoral Dissertation, Idaho State University, Pocatello, 1974.
2. J. E. Holmes, Continued fractions, Math. Teacher, 61 (January 1968) 12–17.
3. I. Niven and H. S. Zukerman, An Introduction to the Theory of Numbers, Wiley, New York, 1960.
4. C. D. Olds, Continued Fractions, New Mathematical Library, Vol. 9, Mathematical Association of America, Washington, D.C., 1963.

Negative Bases with Rational Exponents

David Rudd, *Norfolk State University, Norfolk, VA*

Does $(-1)^{6/2}$ equal $(-1)^3 = -1$, or does $(-1)^{6/2} = \sqrt{(-1)^6} = 1$? And, does $[(-1)^{2/3}]^{3/2}$ equal $(-1)^1 = -1$, or does it equal $\left[\sqrt[3]{(-1)^2} \right]^{3/2} = 1^{3/2}$ $= 1$? Evidently the law of exponents $(a^u)^v = a^{uv}$ need not hold for negative bases with rational exponents. Is it possible, without introducing complex numbers, to develop a simple theory of exponents with negative bases within the real number system? In other words, we want to define a^u ($a < 0$ and u, rational) in such a way that the usual rules

$$\textbf{(i)} \quad a^u \cdot a^v = a^{u+v}, \quad \textbf{(ii)} \quad \frac{a^u}{a^v} = a^{u-v}, \quad \textbf{(iii)} \quad (a^u)^v = a^{uv}$$

hold. These rules, of course, hold for integral exponents. However, it is immediately clear that there is no way to define $a^{1/2}$ as a real number and still satisfy (iii), since this yields

$$0 < \left(a^{1/2}\right)^2 = a^1 = a < 0.$$

Thus, a^u can only be defined for certain rational exponents u.

A rational number will be termed *admissible* if it can be expressed as a rational number with an odd denominator. It is easily verified that the sum, difference, and product of two admissible rationals is an admissible rational. Next, recall that two integers are said *to have the same parity* if they are both even or both odd. Thus, m and n have the same parity if and only if $(-1)^m = (-1)^n$. Evidently $m + n$ and $m - n$ always have the same parity.

The preceding remarks enable us to define

$$a^{p/q} = (-1)^p |a|^{p/q}$$

if q is odd. Note that $a^{p/q}$ is well defined since $p/q = p'/q'$ ($q > q'$ odd) requires that p, p' have the same parity, and so $a^{p/q} = a^{p'/q'}$. According to our definitions,

$$(-1)^{6/2} = (-1)^3 = -1$$

and

$$\left[(-1)^{2/3}\right]^{3/2} = \left[(-1)^2\right]^{3/2} = (1)^{3/2} = (1)^3 = 1.$$

Theorem. *Rules* (i)–(iii) *hold for admissible exponents* u, v.

Proof. For any $u = p/q$ and $v = r/s$ (q, s odd), we have

$$u \pm v = \frac{ps \pm qr}{qs} \quad \text{and} \quad uv = \frac{pr}{qs}.$$

Furthermore, $p \pm r$ and $ps \pm qr$ have the same parity since $p \pm r$ and $ps \pm qr$ are all even if p and r have the same parity, whereas $p \pm r$ and $ps \pm qr$ are all odd if p and r do not have the same parity. It now follows, by definition, that

(i) $$a^u \cdot a^v = (-1)^p |a|^u \cdot (-1)^r |a|^v = (-1)^{p+r} |a|^{u+v}$$
$$= (-1)^{ps+qr} |a|^{u+v} = a^{u+v},$$

and

(ii) $$\frac{a^u}{a^v} = \frac{(-1)^p |a|^u}{(-1)^r |a|^v} = (-1)^{p-r} |a|^{u-v}$$
$$= (-1)^{ps-qr} |a|^{u-r} = a^{u-v}.$$

Similarly, (iii) holds since

$$(a^u)^v = \left[(-1)^p |a|^u \right]^v = \begin{cases} (-1)^r \left| (-1)^p |a|^u \right|^v = (-1)^r |a|^{uv}, & \text{odd } p \\ |a|^{uv}, & \text{even } p \end{cases}$$

agrees with

$$a^{uv} = (-1)^{pr} |a|^{uv} = \begin{cases} (-1)^r |a|^{uv}, & \text{odd } p \\ |a|^{uv}, & \text{even } p \end{cases}.$$

Mathematical Induction and Matrix Inversion

Robert M. Tardiff, *Franklin & Marshall College, Lancaster, PA*

Let $B > 0$ ($B \geqslant 0$) denote the fact that each element b_{ij} of the matrix B satisfies $b_{ij} > 0$ ($b_{ij} \geqslant 0$). Note that $B \geqslant 0$ does not mean $B > 0$ or $B = 0$.

Example. $X = \binom{2}{3} > 0$ and $A = \left(\begin{smallmatrix} 1/2 & 0 \\ 1/2 & 0 \end{smallmatrix} \right) \geqslant 0$. Furthermore,

$$(I - A)X = \begin{pmatrix} \frac{1}{2} & 0 \\ -\frac{1}{2} & 1 \end{pmatrix}\begin{pmatrix} 2 \\ 3 \end{pmatrix} = \begin{pmatrix} 1 \\ 2 \end{pmatrix} > 0 \quad \text{and} \quad (I - A)^{-1} = \begin{pmatrix} 2 & 0 \\ 1 & 1 \end{pmatrix} \geqslant 0.$$

The preceding example illustrates the known result that:

> For a square matrix $A \geqslant 0$, the matrix $I - A$ has an inverse $(I - A)^{-1}$ and $(I - A)^{-1} \geqslant 0$ if $(I - A)X > 0$ for some vector $X > 0$.

We offer an elementary proof of this interesting result. Our proof uses induction only and does not require the knowledge or use of characteristic values of A.

The proof is by mathematical induction on the dimension, $\dim A$, of A. Our result is clearly true when $\dim A = 1$. It remains therefore to show that the result is true for $\dim A = k$ whenever it is true for $\dim A \leqslant k - 1$. To this end, let $r = [k/2]$ be the greatest integer less than or equal to $k/2$. (Note that r and $k - r$ are both less than or equal to $k - 1$.) Now partition the matrix $I - A$ as follows:

$$
\left[
\begin{array}{ccc|ccc}
1 - a_{11} & \cdots & -a_{1r} & -a_{1,r+1} & \cdots & -a_{1k} \\
\vdots & & \vdots & \vdots & & \vdots \\
-a_{r1} & \cdots & 1 - a_{rr} & -a_{r,r+1} & \cdots & -a_{rk} \\
\hline
-a_{r+1,1} & \cdots & -a_{r+1,r} & 1 - a_{r+1,r+1} & \cdots & -a_{r+1,k} \\
\vdots & & \vdots & \vdots & & \vdots \\
-a_{k1} & \cdots & -a_{kr} & -a_{k,r+1} & \cdots & 1 - a_{kk}
\end{array}
\right]
=
\begin{bmatrix}
I - A_{11} & -A_{12} \\
-A_{21} & I - A_{22}
\end{bmatrix}.
$$

Partition the vector X as follows:

$$X = \begin{bmatrix} x_1 \\ \vdots \\ x_r \\ \hline x_{r+1} \\ \vdots \\ x_k \end{bmatrix} = \begin{bmatrix} \vec{x}_1 \\ \vec{x}_2 \end{bmatrix}.$$

Then $(I - A)X > 0$ implies that

$$(I - A_{11})\vec{x}_1 - A_{12}\vec{x}_2 > 0 \qquad (1)$$

and

$$- A_{21}\vec{x}_1 + (I - A_{22})\vec{x}_2 > 0. \qquad (2)$$

Since $A \geqslant 0$, we have $A_{12} \geqslant 0$ and $A_{21} \geqslant 0$. Thus, it follows that

$$(I - A_{11})\vec{x}_1 > 0 \quad \text{and} \quad (I - A_{22})\vec{x}_2 > 0.$$

Hence, by the induction hypothesis,

$$(I - A_{11})^{-1} \geqslant 0 \quad \text{and} \quad (I - A_{22})^{-1} \geqslant 0.$$

Multiplying (1) on the left by $A_{21}(I - A_{11})^{-1}$ (which is greater than or equal to 0) and then adding the result to (2) yields

$$\left(I - \left(A_{22} + A_{21}(I - A_{11})^{-1}A_{12}\right)\right)\vec{x}_2 > 0.$$

Again, by the induction hypothesis,

$$\left(I - \left(A_{22} + A_{21}(I - A_{11})^{-1}A_{12}\right)\right)^{-1} \geqslant 0. \qquad (3)$$

Similarly,

$$\left(I - \left(A_{11} + A_{12}(I - A_{22})^{-1}A_{21}\right)\right)^{-1} \geqslant 0. \qquad (4)$$

Solving

$$\begin{bmatrix} I - A_{11} & - A_{12} \\ - A_{21} & I - A_{22} \end{bmatrix} \begin{bmatrix} B_{11} & B_{12} \\ B_{21} & B_{22} \end{bmatrix} = \begin{bmatrix} I & 0 \\ 0 & I \end{bmatrix}$$

for B_{11}, B_{12}, B_{21}, and B_{22} yields

$$(I - A_{11})B_{11} - A_{12}B_{12} = I \qquad (5)$$
$$(I - A_{11})B_{12} - A_{12}B_{22} = 0 \qquad (6)$$
$$- A_{21}B_{11} + (I - A_{22})B_{21} = 0 \qquad (7)$$
$$- A_{21}B_{12} + (I - A_{22})B_{22} = I. \qquad (8)$$

From (7), we get

$$B_{21} = (I - A_{22})^{-1}A_{21}B_{11}$$

which, substituted into (5), yields

$$B_{11} = \left[I - \left\{ A_{11} + A_{12}(I - A_{22})^{-1}A_{21} \right\} \right]^{-1}.$$

Similarly, (6) can be solved as

$$B_{12} = (I - A_{11})^{-1}A_{12}B_{22},$$

and substitution into (8) gives

$$B_{22} = \left[I - \left\{ A_{22} + A_{21}(I - A_{11})^{-1}A_{12} \right\} \right]^{-1}.$$

It follows, from (3) and (4), that $B_{11} \geqslant 0$ and $B_{22} \geqslant 0$. Since B_{12} and B_{21} are products of non-negative matrices, they too are non-negative. Therefore,

$$(I - A)^{-1} = B = \begin{bmatrix} B_{11} & B_{12} \\ B_{21} & B_{22} \end{bmatrix} \geqslant 0$$

as originally asserted.

Remark. Our method for computing a $k \times k$ inverse $(I - A)^{-1}$ shows how to reduce that problem to that of computing two $r \times r$ inverses and two $(k - r) \times (k - r)$ inverses. While this technique holds for every $r < k$, the value of $r = [k/r]$ seems most appropriate to consider.

2. GEOMETRY

It seems reasonable to say that the earliest development of geometry, as a subject, began with the ancient Egyptians. Their architects used geometric constructions to position and build pyramids, and their priests used geometric principles to assess the taxable value of land areas.

The first substantial change in geometry began when Thales (640–546 B.C.) visited Egypt and returned to Greece with the seeds of this new knowledge. Thales and his student Pythagoras (580–500 B.C.) emphasized rigorous logic in studying the interrelationships among parts of a geometric figure. These seeds evidently bore fruit. Geometry, as a deductive study, flourished in the fertile soil of Greek abstraction, and it was not long before Euclid (330–275 B.C.) appeared on the scene. His great work, *The Elements*, provided a formal and coherent account of the mathematics known in his day.

The most celebrated "unsolved problems" of Euclid's day concerned constructions by compass and straightedge. It was well known, for example, how to construct regular polygons having N sides—but only for values of N equal to $2^n, 3, 5$, and their products. From another perspective, this meant that it was known how to trisect only certain angles. (An angle of 60°, for instance, could not be trisected since it was not known how to divide the circle into $2 \cdot 3^2$ equal parts.) These unsolved problems must have tantalized Greek researchers—most notably Archimedes (287–212 B.C.), whose many great discoveries, including the determination of planar areas, were closely intertwined with properties of regular polygons. Archimedes did provide

27

the first known method for trisecting angles—albeit his technique required a straightedge having two marked points. The answers to these unsolved problems would have to wait . . . more than two thousand years.

The first two articles in this chapter are linked to geometry's earliest concerns: planar areas and compass constructions. In "Pick's Theorem for Simple Closed Polygons," J. B. Harkin develops a surprisingly simple method for determining the area of a simple closed polygon whose vertices lie at lattice points of a grid. This technique, based on a result from dissection theory, is applied to an interesting counting problem. Joe Dan Austin's article "Angle Trisection Using Compass and a Finite Straight-edge" gives an alternative method, to Archimedes' purely geometric tech-nique, for trisecting angles. His article, beginning with a Mascheroni construction*, uses elementary concepts from algebra and trigonometry. Thus, Austin's method of trisection might be viewed as a modern solution to a problem rooted in antiquity.

In 1796, Gauss proved that regular polygons having N sides were constructible with compass and straightedge if and only if N is of the form

$$N = 2^{n_1}(2^{n_2} + 1) \cdots (2^{n_k} + 1),$$

where all n_i are natural numbers and each $2^{n_i} + 1$ is prime.

By now geometry had matured and taken algebra as its partner. Descartes performed the marriage in 1637. The new couple, called coordi-nate geometry, made quite an impact on mathematical thinking: algebraic expressions (such as $x^N = 1$) and geometric interpretations (divide the circle into N equal parts) were now intimately tied together. There were many important offspring, Gauss's result being one of the most famous. Gauss, in fact, was one of the first to use and interpret complex numbers as representing points in the plane.

The next two articles in this chapter focus on this aspect of geometry's development. Geometric figures and algebraic expressions are studied by trigonometric methods. In his article "On Some Properties of the Quadran-gle," Albert Schild uses geometric aspects of complex numbers in order to prove some surprising properties about diagonals of the quadrangle formed by joining the vertices of isosceles right triangles whose hypotenuses are the sides of a given convex quadrangle. These results are also considered for the case of a triangle (viewed as a quadrangle, one of whose sides shrank to zero) and other special quadrangles. Next, Gerald Bergum and Ken Yocom add scope to Pythagoras' famous result. If $\{a, b, c\}$ are integer-sides of a right triangle with hypotenuse c, then DeMoivre's theorem is used to show

---

*In 1797, Mascheroni demonstrated that every problem solvable with compass and straight-edge can also be solved with compass alone. Constructions using only a compass became known as Mascheroni constructions. It may be of interest to note that in 1797, Napoleon Bonaparte (also a geometry buff) became fascinated with the problem: Is there a Mascheroni construction for inscribing a square in a given circle? (Yes!)

that $\{|c^n \cdot \cos n\theta|, |c^n \cdot \sin n\theta|, c^n\}$ are integer-sides of a right triangle with hypotenuse c^n. These sides have c^n as a common factor.

Question: Can one produce relatively prime integer-sides (as, for example, $\{|4a^3 - 3ac^2|, |4b^3 - 3bc^2|, c^3\}$ when $n = 3$)? In "Tchebysheff Polynomials and Primitive Pythagorean Triples," the authors use polynomial expressions for $\cos n\theta$ in order to extend this solution to higher integral powers of c.

In three-dimensional space, regular polyhedra are the natural generalization of regular polygons. In contrast to the existence of many regular polygons, however, there exist only five regular polyhedra. These regular polyhedra (called the Platonic solids) were encompassed in Archimedes' work on semiregular polyhedra (later known as Archimedean solids). Kepler (1571–1630) was the first to verify that there were exactly thirteen Archimedean solids. In "Semiregular Polyhedra," Charles Jepsen derives and classifies these Archimedean solids in terms of the pattern of regular polygons that surround a vertex. Jepsen's article demonstrates that elementary concepts from algebra and geometry still yield interesting and historically significant results.

The preceding articles illustrate the prominent role played by angles in the study and characterization of geometric objects. It may be interesting, therefore, to consider this in an entirely different framework. Frank Harary and David Leep do this in "Angles in Graphs." Having defined graphical distance between points in a connected graph, the authors measure an angle (defined as an ordered triple of distinct points) using the conventional Law of Cosines. Properties of angle measurement defined in this manner are compared and contrasted with the usual properties of angular measurement. In certain graphs, the only angles which occur are 0, $\pi/3$, and π. Such graphs, Harary and Leep show, can be characterized as belonging to one of five classes of connected graphs.

Pick's Theorem for Simple Closed Polygons

J. B. Harkin, *State University of New York, Brockport, NY*

How can one determine the area $A(P)$ enclosed by a simple closed polygon P whose vertices lie at lattice points of a grid?

Figure 1.

One surprisingly simple technique uses the number $B(P)$, of boundary points, and the number $I(P)$, of interior points, of P. This is by Pick's Theorem

$$A(P) = \frac{B(P)}{2} + I(P) - 1.$$

Our purpose here is to develop Pick's Theorem and show how it can be applied to an interesting counting problem.

Theorem 1. *A lattice rectangle R has area*

$$A(R) = \frac{B(R)}{2} + I(R) - 1.$$

where $B(R)$ is the number of lattice points on R, and $I(R)$ is the number of lattice points interior to R.

Proof. Let m and n denote the number of unit lengths in the respective sides of the rectangle R. Then counting the boundary and interior points in terms of m and n, we obtain

$$\frac{B(R)}{2} + I(R) - 1 = \frac{2(m+1) + 2(n-1)}{2} + (n-1)(m-1) - 1$$

$$= m \cdot n$$

$$= A(R).$$

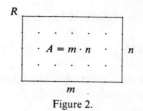

Figure 2.

Corollary. $A(T) = \frac{1}{2} \cdot B(T) + I(T) - 1$ *for a lattice right triangle T.*

Proof. A lattice right triangle T, having legs of m and n units length, can be embedded in a rectangle R having sides of m and n units length (Figure 3).

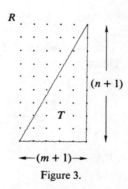

Figure 3.

If B_D denotes the number of lattice points on the hypotenuse of T that are interior to R, then:

$$B(R) = \left[B(T) - B(D) \right] + (m - 1) + n$$

and

$$I(R) = 2I(T) + B(D).$$

Therefore,

$$B(R) + I(R) = B(T) + 2I(T) + (m - 1) + n$$

and

$$B(T) + 2I(T) = B(R) + I(R) - m + 1 - n$$
$$= mn + 2$$
$$= A(R) + 2.$$

Since $A(T) = A(R)/2$, we obtain $A(T) = \frac{1}{2} \cdot B(T) + I(T) - 1$ as asserted.

An extension of the formula $A(T) = \frac{1}{2} \cdot B(T) + I(T) - 1$ to all triangular domains is accomplished through an embedding in a rectangular domain.

Theorem 2. *Every lattice triangle S has area*

$$A(S) = \frac{B(S)}{2} + I(S) - 1.$$

Proof. There are two cases to consider, depending on whether or not one side of S shares lattice points in common with the boundary of the embedding lattice rectangle R.

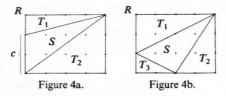

Figure 4a.   Figure 4b.

(i) Consider the case (Figure 4a) where S has a side c, with B_c lattice points, contained in $B(R)$. Clearly,

$$A(S) = A(R) - A(T_1) - A(T_2)$$

$$= \left[\frac{B(R)}{2} + I(R) - 1 \right] - \left[\frac{B(T_1)}{2} + I(T_1) - 1 \right]$$

$$- \left[\frac{B(T_2)}{2} + I(T_2) - 1 \right]$$

$$= \tfrac{1}{2}\left[B(R) - B(T_1) - B(T_2) \right] + \left[I(R) - I(T_1) - I(T_2) \right] + 1.$$

But $B(T_1) + B(T_2) + 2B_c - 2 = B(S) + B(R)$, or

$$B(R) - B(T_1) - B(T_2) = 2B_c - 2 - B(S)$$

and $I(T_1) + I(T_2) + I(S) + [B(S) - B_c - 1] = I(R)$, or

$$I(R) - I(T_1) - I(T_2) = I(S) + B(S) - B_c - 1.$$

Substituting the latter two formulas into the formula for $A(S)$, we obtain

$$A(S) = \tfrac{1}{2}\left[2B_c - 2 - B(S) \right] + \left[I(S) + B(S) - B_c - 1 \right] + 1$$

$$= \frac{B(S)}{2} + I(S) - 1.$$

(ii) The reader can similarly verify that $A(S) = \tfrac{1}{2} \cdot B(S) + I(S) - 1$ for the case (Figure 4b) in which only the vertices of S share lattice points in common with $B(R)$.

Theorem 3. *Let $P = P_1 \cup P_2$ be the lattice polygon formed by lattice polygons P_1 and P_2 which intersect only in a common edge C and have no interior points in common. If $A(P_1) = \tfrac{1}{2} \cdot B(P_1) + I(P_1) - 1$ and $A(P_2) = \tfrac{1}{2} \cdot B(P_2) + I(P_2) - 1$, then $A(P) = \tfrac{1}{2} \cdot B(P) + I(P) - 1$.*

Proof. Let B_C denote the numbers of lattice prints on edge C (Figure 5).

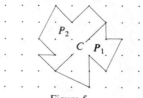

Figure 5.

Then

$$I(P) = I(P_1) + I(P_2) + (B_C - 2)$$

and

$$B(P) = B(P_1) + B(P_2) - 2B_C - 2.$$

Therefore,

$$\begin{aligned}
A(P) &= A(P_1) + A(P_2) \\
&= \left[\frac{B(P_1)}{2} + I(P_1) - 1 \right] + \left[\frac{B(P_2)}{2} + I(P_2) - 1 \right] \\
&= \frac{B(P_1) + B(P_2)}{2} + \left[I(P_1) + I(P_2) \right] - 2 \\
&= \frac{B(P) + 2B_C - 2}{2} + \left[I(P) - B_C + 2 \right] - 2 \\
&= \frac{B(P)}{2} + I(P) - 1.
\end{aligned}$$

Our approach to inductively establish Pick's formula for a lattice polygon P requires a result from dissection theory (see [1]):

Every polygon of n sides can be dissected into $(n - 2)$ triangles whose vertices lie only at the vertices of the polygon.

Triangles whose vertices lie only on lattice points are called *fundamental* (see Figure 6). Figure 7 shows how a polygon of nine sides is dissected into seven fundamental triangles.

Figure 6.   Some fundamental triangles.

Figure 7. Dissection Theorem for $n = 9$.

Theorem 4. *A lattice polygon P has area $A(P) = \frac{1}{2} \cdot B(P) + I(P) - 1$.*

Proof. Let P be a lattice polygon of n sides dissected into $(n - 2)$ triangles. For $n = 3$, Theorem 2 established Pick's formula. Consider $n > 3$. We show that Pick's formula valid for lattice polygons P_n of n sides implies Pick's formula is valid for lattice polygons P_{n+1} of $(n + 1)$ sides. Assume that a lattice polygon P_{n+1} is dissected into a lattice polygon P_n of n sides and a triangle S, where P_n and S have only a common edge C and no common interior points. Then

$$
\begin{aligned}
A(P_{n+1}) &= A(P_n) + A(S) \\
&= \left[\frac{B(P_n)}{2} + I(P_n) - 1 \right] + \left[\frac{B(S)}{2} + I(S) - 1 \right] \\
&= \frac{B(P_n) + B(S)}{2} + \left[I(P_n) + I(S) \right] - 2 \\
&= \frac{B(P_{n+1}) + 2B_C - 2}{2} + \left[I(P_{n+1}) - B_C + 2 \right] - 2 \\
&= \frac{B(P_{n+1})}{2} + I(P_{n+1}) - 1.
\end{aligned}
$$

A complete dissection of a lattice polygon P into fundamental triangles confronts us with a challenge. How do we count the number $\#_f(p)$ of fundamental triangles in a total dissection?

Ordinarily counting problems are among the most difficult problem-solving tasks of mathematics. Theorem 5 provides an interesting illustration of the application of a concept (area), that is essentially characterized by continuity, to a difficult problem of discrete mathematics (counting of fundamental triangles).

Theorem 5. *A lattice polygon P can be dissected into exactly $\#_f(P) = B(P) + 2I(P) - 2$ fundamental triangles.*

Proof. By definition, every fundamental triangle S has $I(S) = 0$ and $B(S) = 3$, and so $A(S) = 1/2$. And $A(P) = \frac{1}{2} \cdot B(P) + I(P) - 1$ for the lattice polygon P. Therefore, $\#_f(P) = A(P)/A(T) = B(P) + 2I(P) - 2$.

Pick's formula can also be derived by reversing the approach taken here. For example (see [2], [3]), Pick's formula can be derived by first introducing and counting the fundamental triangles associated with a complete dissection of the lattice polygon.

REFERENCES

1. H. Eves, A Survey of Geometry, Chapter 5, Dissection Theory, Allyn and Bacon, Boston, 1963, pp. 237–239.
2. J. McNamara and J. Harkin, Genesee on the geoboard, Math. Teachers J., 3 (1975) 61–69.
3. D. DeTemple and J. Robertson, Equivalence of Euler's and Pick's theorems, Math. Teacher, 67 (1974) 222–226.

Angle Trisection Using Compass and Finite Straightedge[†]

Joe Dan Austin, *Rice University, Houston, TX*

The following alternative to Archimedes' method of trisecting angles [see Courant and Robbins, *What is Mathematics?* 1941, p. 138] serves as a nice vehicle for students to reinforce and integrate geometric, trigonometric, and algebraic concepts learned earlier.

We begin by noting that a finite straightedge of length one and a compass can be used to construct an arbitrarily long line segment having markings one unit apart. For example [see W. Page, "Compass constructions of length $\sqrt{n}$," *MATYCJ*, 10 (1976) 180–181], suppose P_0 and P_1 (Figure 1) are one unit apart. Keeping the compass fixed at one unit length, construct a circle with center P_1 and mark off arcs $\overset{\frown}{P_0 A}$, $\overset{\frown}{AB}$, and $\overset{\frown}{BP_2}$ on this circle. Since triangles $P_0 A P_1$, $P_1 AB$, and $BP_1 P_2$ are equilateral, points $\{P_0, P_1, P_2\}$ are collinear and $\overline{P_0 P_2}$ has length two units. The finite straightedge can then be used to draw the line segments $\overline{P_0 P_1}$ and $\overline{P_1 P_2}$, and thus the line segment $\overline{P_0 P_2}$. Proceeding in a similar fashion, one obtains collinear points $\{P_0, P_1, P_2, \ldots, P_n\}$ such that $\overline{P_0 P_n}$ has length n units.

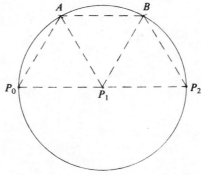

Figure 1.

[†]A finite straightedge supplemented by a compass becomes a "ruler" and so this is *not* the classical trisection problem.

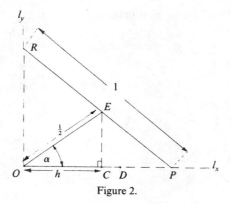

Figure 2.

Suppose (Figure 2) that acute $\sphericalangle EOD$, having measure α, is to be trisected. Also assume that side $\overline{OE}$ has length $1/2$, which we obtain by bisecting the unit length. By the previous remarks, our finite straightedge of length one can be used to extend $\overline{OD}$ to a line l_x. Now construct $\overline{EC}$ perpendicular to l_x, and let h denote the length of $\overline{OC}$. Next, construct line l_y perpendicular to $\overline{OD}$ at O. It is now possible to place the straightedge through point E so that its end points intersect lines l_x and l_y at P and R, respectively. To be sure, let x be the length of $\overline{OP}$ and let y be the length of $\overline{OR}$. Since triangles ECP and ROP are similar, one can substitute

$$y = \frac{x\sqrt{(1/4) - h^2}}{x - h}$$

into $x^2 + y^2 = 1$ in order to obtain

$$x^4 - 2hx^3 - \tfrac{3}{4}x^2 + 2hx - h^2 = 0. \qquad (*)$$

The solutions $x > h$ of $(*)$ determine where the finite straightedge placed through E will intersect the lines l_x and l_y (as shown in Figure 2). One solution is $x = 2h$. Therefore $(*)$ can be rewritten

$$(x - 2h)\left(x^3 - \frac{3}{4}x + \frac{h}{2}\right) = 0.$$

There are three distinct real roots of $x^3 - 3x/4 + h/2 = 0$. This follows because the polynomial is negative for $x = -2$, positive for $x = 0$, and negative for $x = 1/2$ (since $0 < \alpha < 90°$ implies that $h < 1/2$). The roots of the general equation

$$x^3 + ax + b = 0 \qquad (**)$$

(all cubic polynomials can be reduced to this form by a linear transformation) having three distinct real roots, can be expressed in trigonometric form (see [1], for example) by first finding Φ such that

$$\cos \Phi = \frac{-b/2}{\sqrt{-a^3/27}} \qquad (a < 0).$$

Then the three distinct, real roots of (**) are

$$2\sqrt{-\frac{a}{3}} \cos \frac{\Phi}{3}, \quad 2\sqrt{-\frac{a}{3}} \cos\left(\frac{\Phi}{3} + 120°\right),$$

and

$$2\sqrt{-\frac{a}{3}} \cos\left(\frac{\Phi}{3} + 240°\right).$$

For the particular equation $x^3 - 3x/4 + h/2 = 0$, we have $\cos \Phi = -2h$. As $\cos \alpha = 2h$ (Figure 2), $\Phi = 180° - \alpha$. Therefore, the three roots are

$$r_1 = \cos\left(60° - \frac{\alpha}{3}\right), \quad r_2 = \cos\left(180° - \frac{\alpha}{3}\right), \quad \text{and} \quad r_3 = \cos\left(300° - \frac{\alpha}{3}\right).$$

Since $0 < \alpha < 90°$, we have

$$r_2 < 0 < r_3 < \frac{1}{2} < r_1 < \frac{\sqrt{3}}{2}.$$

Therefore, there are either two or three ways (when $h < r_3$ and $2h \neq r_1, r_3$) to positive $\overline{PR}$ through E. For the trisection given here, we want the placement with $x = r_1$ so that $\overline{OP} = (60° - \alpha/3)$. There are two possibilities:

(i) If $r_1 = 2h$, then $\cos(60° - \alpha/3) = \cos \alpha$, and so $\alpha = 45°$. This angle is easily trisected by bisecting a 60° angle (constructed, for example, as $\angle AP_0P_1$ in Figure 1) and then bisecting one of the resulting 30° angles.

(ii) For $r_1 \neq 2h$ (i.e., $\alpha \neq 45°$), construct a line l perpendicular to l_x at P (Figure 3), and draw a circle with center O and radius one. This circle intersects line l (since $r_1 < 1$) at some point V. Then $\angle VOP$ has measure

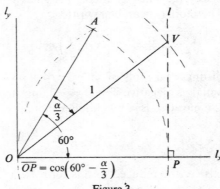

Figure 3.

$60° - \alpha/3$, and an angle with measure $\alpha/3$ is obtained by constructing the $60°$ angle $\angle APO$. Specifically, $\angle AOV$ is the desired angle with measure $\alpha/3$.

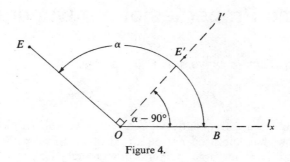

Figure 4.

Finally, consider the case where $90° < \alpha < 180°$. Construct line l' perpendicular to $\overline{EO}$ (Figure 4), and mark off E' on l' so that $\overline{OE'} = 1$. The acute angle $E'OB$ of measure $\alpha - 90°$ can be trisected, as before, to give an angle measuring $\alpha/3 - 30°$. This angle, added to a $30°$ angle, yields an angle of measure $\alpha/3$.

REFERENCE
1. C. R. C. Standard Mathematical Tables, C. D. Hodgman (Editor), Chem. Rubber Publ. Co., Cleveland, 1959, pp. 358–359.

On Some Properties of the Quadrangle

Albert Schild, *Temple University, Philadelphia, PA*

Complex numbers are frequently viewed with awe and suspicion by students. To show that one need not be afraid of them, we will use them to prove some rather remarkable theorems in Geometry.

In "A Survey of Geometry" [1] we find the following startling result:

Theorem 1. *If right isosceles triangles ABA', BCB', CDC', DAD' are described exteriorly on the sides as hypotenuses of a convex quadrangle ABCD, then A'C' is equal and perpendicular to B'D' (Figure 1).*

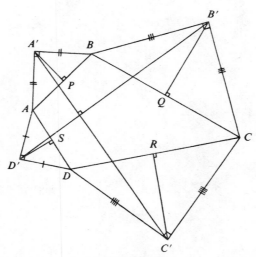

Figure 1.

This theorem appears as a special case of V. Thébault's problem [2]. Our objective is to use an iterative procedure in order to extend this theorem. Thus, we take A', B', C', and D' as the vertices of a new quadrangle and repeat the construction of the theorem. This gives rise to a new quadrangle

$A''B''C''D''$, and this procedure is repeated with the new quadrangle, etc. Properties of this sequence of quadrangles will be investigated.

For the sake of completeness we give a proof of the theorem mentioned above. In the proof, we shall make use of the following geometric properties of complex numbers.

If a and b are two points in the complex plane, then:

(i) $|b - a|$ represents the length of the line segment joining a to b.

(ii) arg $(b - a)$ is the angle which this line segment makes with the real axis.

(iii) arg $(a \cdot b) = \arg a + \arg b$. In particular, $\arg(a \cdot i) = \arg a + \pi/2$.

(iv) Let a, b, c be three points in the complex plane such that $|b - a| = |c - a|$. Then $c - a = i(b - a)$ if $\sphericalangle cab = \pi/2$ and $c - a = -i(b - a)$ if $\sphericalangle cab = -\pi/2$.

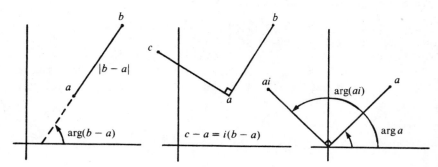

Figure 2.

Proof of Theorem 1. Let the vertices A, B, C, D of the quadrangle (Figure 1) be denoted by the complex numbers a, b, c, d. Let P, Q, R, S be the midpoints of the sides AB, BC, CD, DA, respectively. Then P, Q, R and S have the coordinates: $(a + b)/2$, $(b + c)/2$, $(c + d)/2$, $(d + a)/2$. Since

$$a' - \left(\frac{a + b}{2}\right) = i\left(b - \left(\frac{a + b}{2}\right)\right),$$

the point A will have coordinate

$$a' = \frac{a + b}{2} + i\left(b - \frac{a + b}{2}\right) = \frac{a + b}{2} + i\left(\frac{b - a}{2}\right) = \left(\frac{1 - i}{2}\right)(a + ib).$$

Similarly, for the coordinates of B', C', D' we get:

$$b' = \left(\frac{1 - i}{2}\right)(b + ic), \quad c' = \left(\frac{1 - i}{2}\right)(c + id), \quad d' = \left(\frac{1 - i}{2}\right)(d + ia).$$

From this we obtain:

$$a' - c' = \left(\frac{1-i}{2}\right)(a + ib) - \left(\frac{1-i}{2}\right)(c + id)$$

$$= \left(\frac{1-i}{2}\right)\{(a - c) + i(b - d)\}$$

$$b' - d' = \left(\frac{1-i}{2}\right)(b + ic) - \left(\frac{1-i}{2}\right)(d + ia)$$

$$= -i\left(\frac{1-i}{2}\right)\{(a - c) + i(b - d)\} = -i(a' - c').$$

Therefore, $|b' - d'| = |-i(a' - c')| = |a' - c'|$ and $A'C' = B'D'$. Furthermore, $\arg(b' - d') = -\pi/2 + \arg(a' - c')$ shows that $A'C' \perp B'D'$.

Remark. Theorem 1 is also true if the right isosceles triangles are described internally on the sides of $ABCD$. Readers can also verify that the theorem is true for nonconvex quadrangles.

We consider now A', B', C', and D' (Figure 1) as vertices of a new quadrangle and perform the same construction. That is, we describe right isosceles triangles $A'B'A''$, $B'C'B''$, $C'D'C''$, $D'A'D''$ exteriorly on the sides of quadrangle $A'B'C'D'$ as hypotenuses. Let the coordinates of A'', B'', C'', D'' be denoted a'', b'', c'', d'', respectively. Then, as in Theorem 1,

$$a'' = \left(\frac{1-i}{2}\right)(a' + ib') \qquad c'' = \left(\frac{1-i}{2}\right)(c' + id')$$

$$b'' = \left(\frac{1-i}{2}\right)(b' + ic') \qquad d'' = \left(\frac{1-i}{2}\right)(d' + ia')$$

and

$$a'' - c'' = \left(\frac{1-i}{2}\right)\left[a' - c' + i(b' - d')\right]$$

$$b'' - d'' = -i\left(\frac{1-i}{2}\right)\left[a' - c' + i(b' - d')\right] = -i(a'' - c''),$$

so that $A''C'' = B''D''$ and $A''C'' \perp B''D''$. But how are the diagonals $A''C''$ and $B''D''$ related to $A'C'$ and $B'D'$?

Theorem 2. *For the quadrangle $A''B''C''D''$, we have*
  (i) *The point of intersection of the diagonals $A''C''$ and $B''D''$ coincides with the point of intersection of the diagonals $A'C'$ and $B'D'$.*
  (ii) *$A''C'' = B''D'' = \sqrt{2} \cdot A'C'$.*
  (iii) *The diagonals $A''C''$ and $B''D''$ intersect the diagonals $A'C'$ and $B'D'$ at an angle of $\pi/4$.*

Proof. Since the diagonals $A'C'$ and $B'D'$ of quadrangle $A'B'C'D'$ intersect at right angles, we may place this quadrangle in the complex plane so that the diagonals lie along the real and imaginary axes and, therefore, intersect at the origin (Figure 3).

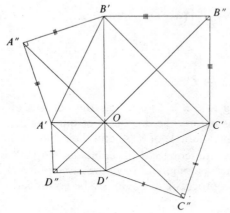

Figure 3.

In this new coordinate system, a' and c' are real, and b' and d' are pure imaginary. For convenience let

$$a' = \alpha', \quad c' = \gamma', \quad b' = i\beta', \quad d' = i\delta',$$

where α', β', γ', δ' are real. Note that $\gamma' - \alpha' = \beta' - \delta'$. We can now express the points a'', b'', c'', d'' in terms of α', β', γ', δ' as

$$a'' = \left(\frac{1-i}{2}\right)(\alpha' - \beta'), \qquad c'' = \left(\frac{1-i}{2}\right)(\gamma' - \delta'),$$

$$b'' = i\left(\frac{1-i}{2}\right)(\beta' + \gamma'), \qquad d'' = i\left(\frac{1-i}{2}\right)(\delta' + \alpha').$$

Furthermore,

$\arg a'' = \arg((1-i)/2) + \arg(\alpha' - \beta') = -(\pi/4) + \pi = 3\pi/4,$
$\arg b'' = \arg i + \arg((1-i)/2) + \arg(\beta' + \gamma') = (\pi/2) - (\pi/4) + 0$
$\quad\quad = \pi/4,$
$\arg c'' = \arg((1-i)/2) + \arg(\gamma' - \delta') = -(\pi/4) + 0 = -(\pi/4),$
$\arg d'' = \arg i + \arg((1-i)/2) + \arg(\gamma' + \alpha') = (\pi/2) - (\pi/4) + \pi$
$\quad\quad = -(3\pi/4).$

Thus, B'' and D'' lie on the angle bisector of the first and third quadrants, and A'' and C'' lie on the angle bisector of the second and fourth quadrants. It follows, from this, that $A''C''$ and $B''D''$ pass through the origin and make angles of $\pi/4$ and $3\pi/4$ with diagonals $A'C'$ and $B'D'$.

The assertion that $A''C'' = B''D'' = \sqrt{2} \cdot A'C'$ is also clear since

$$A''C'' = |a'' - c''| = \left| \frac{1-i}{2} \left[\alpha' - \beta' - \gamma' + \delta' \right] \right|$$

$$= \left| \frac{1-i}{2} \left[(\alpha' - \gamma') + (\delta' - \beta') \right] \right|$$

$$= \left| \frac{1-i}{2} \cdot 2(\alpha' - \gamma') \right| = |(1-i)||\alpha' - \gamma'| = \sqrt{2}\,(\gamma' - \alpha')$$

$$= \sqrt{2} \cdot A'C'.$$

Remark. Taking A'', B'', C'', D'' as the vertices of a new quadrangle and proceeding as before, the diagonals of quadrangle $A'''B'''C'''D'''$ will again be rotated by an angle of $\pi/4$, and their lengths will be multiplied by $\sqrt{2}$. Hence, these diagonals will coincide with those of quadrangle $A'B'C'D'$, be twice as long, and have the same point of intersection as the original ones. This process can obviously be continued with corresponding results.

Some Special Quadrangles. Let us briefly apply Theorem 1 to some "special" quadrangles. For instance, we can view a triangle as a "special" quadrangle, one of whose sides shrank to zero. Or we may look at a "special" quadrangle having three vertices collinear. This quadrangle may degenerate even further by squashing it so that its area becomes zero (i.e., the quadrangle may have all its vertices collinear). We invite readers to verify the results below for these "special" quadrangles.

(a) Consider the isosceles right triangles described exteriorly on the sides of $\triangle ABC$ shown below. An application of Theorem 1 to the quadrangle $ABCD$, where D coincides with C (therefore D' coincides with C'), yields $A'C = B'C'$ and $A'C \perp B'C'$.

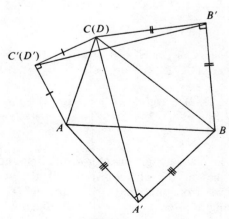

Figure 4.

Remark.   Two other forms of Theorem 1, applied to this special quadrangle, can be obtained by cyclically permuting the letters A, B, C—namely, $B'A' = C'A'$ and $B'A \perp C'A'$, and $C'B = A'B'$ and $C'B \perp A'B'$.

(b)   Let B be a point on the side AC of $\triangle ABC$ (Figure 5), and let isosceles right triangles be described exteriorly with AB, BC, CD, DA as hypotenuses. Then Theorem 1, applied to $ABCD$, yields $A'C' = B'D'$ and $A'C' \perp B'D'$.

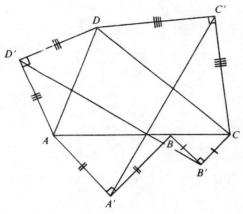

Figure 5.

(c)   Let A, B, C, D be collinear points with B and D between A and C (Figure 6), and let the corresponding isosceles right triangles, with hypotenuses along AC, have vertices A', B' on one side of AC and vertices C', D' on the other. Then Theorem 1 (applied to quadrangle $ABCD$) yields $A'C' = B'D'$ and $A'C' \perp B'D'$.

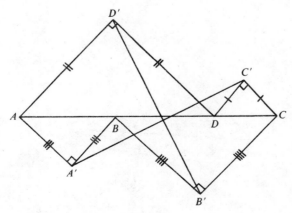

Figure 6.

(d)   Let A, B, C be collinear points, with B between A and C (Figure 7), and let the isosceles right triangles with hypotenuses AB, BC, AC have vertex A' on one side of AC, and vertices B', C' on the other side of AC. Then Theorem 1 (applied to quadrangle $ABCD$, where D coincides with A) yields $AC' = A'B'$ and $AC' \perp A'B'$.

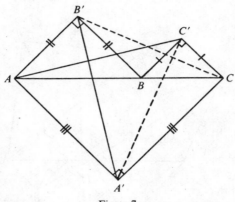

Figure 7.

Remark.   Two other forms of Theorem 1, applied to this special quadrangle, can be obtained by cyclically permuting the letters A, B, C—namely, $BA' = B'C'$ and BA' (extended) $\perp B'C'$, and $CB' = C'A'$ and $CB' \perp C'A'$.

(e)   Let A, B, C, D be collinear points, in that order (Figure 8), and let the isosceles right triangles with hypotenuses along AD be described so that vertices A', B', C' are on one side of AD, and D' is on the other. Then $A'C' = B'D'$ and $A'C' \perp B'D'$.

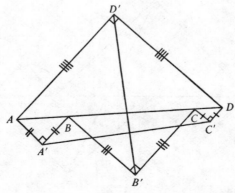

Figure 8.

(f)   Let $ABCD$ and $A'B'C'D'$ be two squares, both oriented clockwise or both oriented counterclockwise, having a common vertex at A (Figure 9). Then $BB' = DD'$ and $BB' \perp DD'$. [Consider the special quadrangle $CAC'A$, and apply Theorem 1 to the vertices B, D, C', D' of the isosceles right triangles having CA, AC', $C'A$, and AC as hypotenuses.]

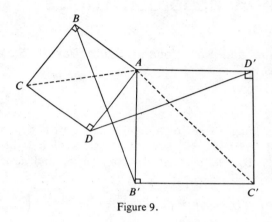

Figure 9.

REFERENCES

1.   Howard Eves, A Survey of Geometry, vol. 2, Allyn and Bacon, Boston, 1965, p. 183.
2.   V. Thébault, Amer. Math. Monthly, 50 (Aug–Sept 1943) 459.

Tchebysheff Polynomials and Primitive Pythagorean Triples

Gerald Bergum and Ken Yocom, *South Dakota State University, Brookings, SD*

If $\{a,b,c\}$ are integer sides of a right triangle with hypotenuse c, then $\{|c^n \cdot \cos n\theta|, |c^n \cdot \sin n\theta|, c^n\}$ are integer sides of a right triangle with hypotenuse c^n. To see this, let $\theta = \cos^{-1}(a/c)$ and use De Moivre's theorem:

$$(a + ib)^n = (c \cdot \cos\theta + ic \cdot \sin\theta)^n = c^n \cdot \cos n\theta + ic^n \cdot \sin n\theta.$$

Since a, b, c, are integers, so also are c^n and

$$c^n \cdot \cos n\theta = \text{Real}\{(a + ib)^n\} \quad \text{and} \quad c^n \cdot \sin n\theta = \text{Imag}\{(a + ib)^n\}.$$

Three numbers $\{x, y, z\}$ will be called primitive if their largest common factor equals one; we denote this $(x, y, z) = 1$. For example, $\{|c^n \cdot \cos n\theta|, |c^n \cdot \sin n\theta|, c^n\}$ is non-primitive since

$$(|c^n \cdot \cos n\theta|, |c^n \cdot \sin n\theta|, c^n) = c^n.$$

Can we produce primitive integer-sides of a right triangle with hypotenuse c^n when the set of integers $\{a, b, c\}$ is primitive? An easily verified solution for $n = 3$ is given by

$$\{|4a^3 - 3ac^2|, |4b^3 - 3bc^2|, c^3\}.$$

The purpose of this note is to show how to extend the above solution to higher integral powers of c. Our objective is to prove, and illustrate the following:

Theorem. *Let $\{a,b,c\}$ be primitive, and suppose n is a positive integer.*
(i) *If n is odd, there exists an nth degree polynomial $T_n(x)$ such that*

$$\left\{|c^n T_n\left(\frac{a}{c}\right)|, |c^n T_n\left(\frac{b}{c}\right)|, c^n\right\}$$

is primitive;

48

(ii) *If $n = 2^m \cdot t$ (t, odd), there exist polynomials $T_{n/2^i}(x)$ (i = 1, 2, ..., m) of degree $n/2^i$ such that*

$$\left\{ \left| c^n T_n\left(\frac{a}{c}\right) \right|, \left| 2^m c^n \prod_{i=1}^{m} T_{n/2^i}\left(\frac{a}{c}\right) \cdot T_{n/2^m}\left(\frac{b}{c}\right) \right|, c^n \right\}$$

is primitive.

Proof. Both assertions are based on the well-known identity (see [2], for example) expressing $\cos n\theta$ as a Tschebysheff polynomial

$$\cos n\theta = \sum_{k=0}^{m} (-1)^k \frac{n}{n-k} \binom{n-k}{n} 2^{n-2k-1} \cos^{n-2k}\theta,$$

where

$$m = \begin{cases} \dfrac{n}{2}, & \text{for even } n, \\[2mm] \dfrac{n-1}{2}, & \text{for odd } n. \end{cases}$$

If we write $T_n(x) = \cos n\theta$ when $x = \cos\theta$, the above expression reads

$$T_n(x) = \sum_{k=0}^{m} (-1)^k \frac{n}{n-k} \binom{n-k}{k} 2^{n-2k-1} x^{n-2k}. \tag{1}$$

This definition of T_n also yields

$$T_n(\sin\theta) = \cos n\left(\frac{\pi}{2} - \theta\right) = \begin{cases} \cos n\theta, & n = 4t, \\ \sin n\theta, & n = 4t+1, \\ -\cos n\theta, & n = 4t+2, \\ -\sin n\theta, & n = 4t+3. \end{cases}$$

We now prove (i) and (ii) separately.

(i) It follows, from $T_n(a/c) = \cos n\theta$ and $T_n(b/c) = \pm\sin n\theta$, that $\{c^n T_n(a/c)\}^2 + \{c^n T_n(b/c)\}^2 = \{c^n\}^2$. Since binomial coefficients are integral,

$$\frac{n}{n-k} \binom{n-k}{k} = \binom{n-k}{k} + \binom{n-k-1}{k-1}$$

is integral. Therefore, (1) establishes that

$$c^n T_n\left(\frac{a}{c}\right) = \sum_{k=0}^{m} (-1)^k \frac{n}{n-k} \binom{n-k}{k} 2^{n-2k-1} \cdot a^{n-2k} \cdot c^{2k} \tag{2}$$

and

$$c^n T_n\left(\frac{b}{c}\right) = \sum_{k=0}^{m} (-1)^k \frac{n}{n-k} \binom{n-k}{k} 2^{n-2k-1} \cdot b^{n-2k} \cdot c^{2k} \tag{3}$$

are integers. It remains only to verify that $\{|c^n T_n(a/c)|, |c^n T_n(b/c)|, c^n\}$ is primitive. Suppose $p > 2$ is a prime which divides each term of this triple. Clearly, p divides c. Since p divides each summand of (2) when $k \geq 1$, it also divides the term $2^{n-1}a^n$ when $k = 0$. Therefore p divides a. Similarly, (3) establishes that p divides b. Since $\{a,b,c\}$ was primitive, we have a contradiction. Thus, $\{|c^nT_n(a/c)|, |c^nT_n(b/c)|, c^n\}$ is primitive.

(ii) If $n = 2t$ (t, odd), then

$$\sqrt{c^{2n} - c^{2n}T_n^2(a/c)} = \pm c^n\sin 2t\theta = \pm 2c^n\sin t\theta \cdot \cos t\theta$$

$$= \pm 2c^nT_{n/2}\left(\frac{b}{c}\right)T_{n/2}\left(\frac{a}{c}\right).$$

For $n = 2^2t$, we obtain

$$\sqrt{c^{2n} - c^{2n}T_n^2\left(\frac{a}{c}\right)} = \pm 2c^n\sin 2t\theta \cdot \cos 2t\theta$$

$$= \pm 2^2\sin t\theta \cdot \cos t\theta \cdot \cos 2t\theta$$

$$= \pm 2^2c^nT_{n/4}\left(\frac{b}{c}\right)T_{n/4}\left(\frac{a}{c}\right)T_{n/2}\left(\frac{a}{c}\right).$$

A simple inductive argument shows that for $n = 2^mt$ ($m > 2$),

$$\sqrt{c^{2n} - c^{2n}T_n^2\left(\frac{a}{c}\right)} = \pm 2^mc^n\prod_{i=1}^{m} T_{n/2^i}\left(\frac{a}{c}\right)T_{n/2^m}\left(\frac{b}{c}\right).$$

As in (i), the triple

$$\left\{|c^nT_n(a/c)|, \left|2^mc^n\prod_{i=1}^{m} T_{n/2^i}\left(\frac{a}{c}\right)T_{n/2^m}\left(\frac{b}{c}\right)\right|, c^n\right\}$$

is primitive.

Example.   Let $\{a,b,c\}$ be primitive and consider $n = 8$. By (1),

$$T_1(x) = x$$
$$T_2(x) = 2x^2 - 1$$
$$T_4(x) = 8x^4 - 8x^2 + 1$$
$$T_8(x) = 128x^8 - 256x^6 + 160x^4 - 32x^2 + 1.$$

Therefore,

$$\{|128a^8 - 256a^6c^2 + 160a^4c^4 - 32a^2c^6 + c^8|,$$

$$|8(8a^4 - 8a^2c^2 + c^4)(2a^2 - c^2)ab|, c^8\}$$

is the corresponding primitive triple.

REFERENCES
1. N. Schaumberger, Problem 75, TYCMJ, 7 (1976) 33.
2. W. J. Wagner, Two explicit expressions for cos ny, Math. Teacher, 67 (1974) 234–237.

Semiregular Polyhedra

Charles H. Jepsen, *Grinnell College, Grinnell, IA*

Information about the regular polyhedra is fairly accessible. The semiregular polyhedra are also known, although discussions are seldom carried out in enough detail to be well understood by students with an elementary mathematics background. Accordingly, we shall derive the semiregular polyhedra using only concepts from geometry and elementary algebra. In so doing, we demonstrate the important point that basic mathematics can yield interesting and significant results.

We begin by collecting the terminology and results needed to derive an essential inequality (1). Our classification of the semiregular polyhedra will follow directly from the solutions to this inequality.

A polyhedron is a solid whose faces are polygons. The intersections of the faces are called *edges*, and the points where the edges meet are called *vertices*. We shall consider only polyhedra with regular polygons as faces.

A polygon is *regular* if all its sides are of equal length and all its vertex angles are equal. A regular polygon with n sides is called an *n-gon*. Since a polygon of n sides can be partitioned into $(n - 2)$ nonoverlapping triangles, the vertex angle of an n-gon is $(n - 2)\pi/n$. (See Figure 1, for example.)

$$\frac{3\pi}{5}$$

Figure 1.

A polyhedron is said to be *convex* if it contains the line segment joining any two of its points. (This is equivalent to saying that the polyhedron lies entirely on one side of any plane containing one of its faces.)* Note that a

---

*A polygon is said to be convex if it contains the line segment joining any two of its points. (This is equivalent to saying that the polygon lies entirely on one side of any line containing one of the sides.) Note that convex polyhedra are the natural generalization of convex polygons.

vertex of a convex polyhedron must "stick out" (Figure 2). We shall need the following consequence of this:

The sum of the vertex angles of the polygons surrounding a vertex of a convex polyhedron is less than 2π.

This is geometrically obvious since one can always lay the surrounding faces, joined at their common vertex, on a flat surface.

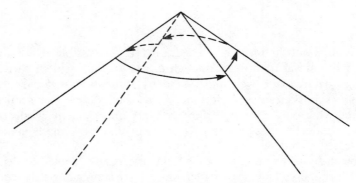

The total angle at a vertex of a convex polyhedron is less than 2π.

Figure 2.

Before defining semiregular polyhedra, we briefly consider the more familiar regular polyhedra. A polyhedron is *regular* if its faces are congruent regular polygons and each vertex has the same number of faces surrounding it. That is, a regular polyhedron has all its faces alike and all its vertices alike. As is well known, there are only five regular polyhedra (Figure 3).

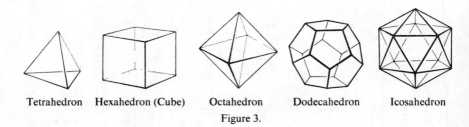

Tetrahedron   Hexahedron (Cube)   Octahedron   Dodecahedron   Icosahedron

Figure 3.

Suppose we attempt to give a corresponding "local" definition for semiregularity of polyhedra. For example, we might call a polyhedron semiregular if its faces are regular polygons and each vertex has the same pattern of faces surrounding it. That is, a semiregular polyhedron has all its vertices alike, but its faces may be different. However, the polyhedron in

Figure 4 becomes semiregular under such a definition. Yet, this solid is not sufficiently uniform in appearance to be called semiregular by most geometers. Consequently, we shall use a "global" definition which excludes this solid from the collection of semiregular polyhedra.

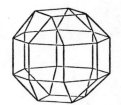

Pseudo-rhombicuboctahedron
Figure 4.

Our definition of semiregularity, which depends on symmetry, is essentially that of Ball and Coxeter ([1], 136). By a *symmetry* of a solid, we mean a succession of rotations and reflections of the solid in three-dimensional space, which leaves the solid unchanged in appearance.

We define a polyhedron to be *semiregular* if its faces are regular polygons and, given any two vertices, there is a symmetry of the polyhedron which sends one vertex to the other. That is, the global appearance of the polyhedron is the same when viewed from any vertex. It follows that a semiregular polyhedron is completely specified by giving the pattern of regular polygons that surround a vertex. If we encounter an n_1-gon, followed by an n_2-gon, followed by an n_3-gon, . . . , followed by an n_r-gon as we cycle around a vertex, we say that the semiregular polyhedron has vertex symbol $(n_1, n_2, \ldots, n_r)$. For example, the cube (Figure 5a) is denoted $(4, 4, 4)$ and the cuboctahedron (Figure 5b) is denoted $(3, 4, 3, 4)$.

Figure 5a.   Figure 5b.

Consider a convex semiregular polyhedron with vertex symbol $(n_1, n_2, \ldots, n_r)$. It will be convenient to let $N_1, N_2, \ldots, N_r$ denote the n_i arranged in increasing order. Thus, $3 \leqslant N_1 \leqslant N_2 \leqslant \cdots \leqslant N_r$ (see, for example, Figure 6).

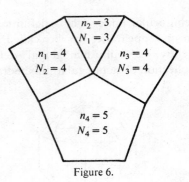

Figure 6.

As was noted earlier,

$$\frac{(n_1 - 2)\pi}{n_1} + \frac{(n_2 - 2)\pi}{n_2} + \cdots + \frac{(n_r - 2)\pi}{n_r} < 2\pi.$$

This can be simplified and rewritten

$$\frac{r}{2} - 1 < \frac{1}{n_1} + \frac{1}{n_2} + \cdots + \frac{1}{n_r}.$$

Since the set $\{N_1, N_2, \ldots, N_r\}$ is simply a reordering of the set $\{n_1, n_2, \ldots, n_r\}$, we also have

$$\frac{r}{2} - 1 < \frac{1}{N_1} + \frac{1}{N_2} + \cdots + \frac{1}{N_r}. \tag{1}$$

Since each $N_i \geqslant 3$, the right-hand side of (1) is less than or equal to $r/3$. Therefore, it follows from (1) that each vertex is surrounded by at most five faces. Indeed, $r \geqslant 6$ implies that

$$\frac{1}{N_1} + \frac{1}{N_2} + \cdots + \frac{1}{N_r} \leqslant \frac{r}{3} \leqslant \frac{r}{2} - 1,$$

and this contradicts (1). Thus, r is either 3, 4, or 5. We shall consider each case separately.

Case $r = 5$. Here (1) becomes

$$\frac{3}{2} < \frac{1}{N_1} + \frac{1}{N_2} + \frac{1}{N_3} + \frac{1}{N_4} + \frac{1}{N_5}. \tag{2}$$

Since $N_1 \leqslant N_2 \leqslant N_3 \leqslant N_4 \leqslant N_5$, inequality (2) forces $N_1 = N_2 = N_3 = N_4 = 3$. For if $N_4 \geqslant 4$, then

$$\frac{1}{N_1} + \frac{1}{N_2} + \frac{1}{N_3} + \frac{1}{N_4} + \frac{1}{N_5} \leqslant \frac{1}{3} + \frac{1}{3} + \frac{1}{3} + \frac{1}{4} + \frac{1}{4} = \frac{3}{2}.$$

Thus $N_1 = N_2 = N_3 = N_4 = 3$ and we see, from $3/2 < 4(1/3) + 1/N_5$, that $N_5 \leqslant 5$. (This type of argument occurs repeatedly. The reader is invited to fill in the details from now on.) The preceding remarks show that there are only three possible vertex symbols

$$(3,3,3,3,3)$$
$$(3,3,3,3,4)$$
$$(3,3,3,3,5)$$

associated with five faces surrounding a vertex.

Case r = 4.   Here (1) becomes

$$1 < \frac{1}{N_1} + \frac{1}{N_2} + \frac{1}{N_3} + \frac{1}{N_4}. \tag{3}$$

This requires that $N_1 = 3$ and $N_2 \leqslant 4$.

Neither (3, 4, 4, 5) nor (4, 4, 3, 5) is possible at this vertex.

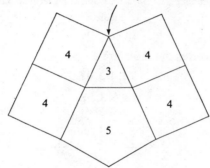

Figure 7.

(a) Suppose $N_2 = 4$. Then (3) gives $N_3 = 4$ and $N_4 \leqslant 5$. Figure 7 shows that the vertex symbols $(3,4,4,5)$ and $(4,4,3,5)$ are impossible. (Here and in subsequent figures the circled vertex is the starting point of the drawing.) The remaining possible vertex symbols are

$$(3,4,4,4)$$
$$(4,3,4,5).$$

(b) Suppose $N_2 = 3$. Then $N_3 \leqslant 5$. If $N_3 = 3$, then $\{3,3,3,N_4\}$ satisfies (3) for any $N_4 \geqslant 3$. Thus, a possible vertex symbol is

$$(3,3,3,m) \quad \text{for any} \quad m \geqslant 3.$$

If $N_3 > 3$, we have the solution set $\{3,3,N_3,N_4\}$ and the possible vertex symbols $(3,3,k,m)$ and $(3,k,3,m)$ that can be made from this set. Since k and m are bigger than 3, Figure 8 shows that $(3,3,k,m)$ is not possible. And Figure 9 shows that $k = m$ for $(3,k,3,m)$.

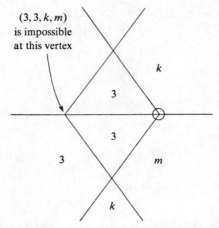

The vertex symbol $(3, 3, k, m)$ with k and m bigger than 3 is impossible.

Figure 8.

Therefore, possible vertex symbols are

$$(3, 4, 3, 4)$$

$$(3, 5, 3, 5).$$

Case $r = 3$.   Here (1) becomes

$$\frac{1}{2} < \frac{1}{N_1} + \frac{1}{N_2} + \frac{1}{N_3}. \tag{4}$$

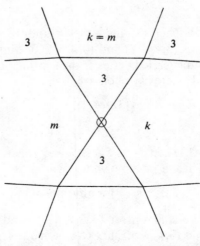

The vertex symbol $(3, k, 3, m)$ with k and m bigger than 3 implies $k = m$.

Figure 9.

(a) Suppose $N_1 = 5$. Then $N_2 \leqslant 6$ and $\{N_1, N_2, N_3\}$ satisfies (4) for $N_3 \geqslant N_2$, and $N_2 = 5$ or 6. If we consider the vertex symbol $(5, k, m)$, Figure 10 shows that $k = m$.

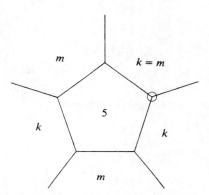

The vertex symbol $(5, k, m)$ gives $k = m$.

Figure 10.

Therefore, possible vertex symbols are

$$(5, 5, 5)$$
$$(5, 6, 6).$$

(b) Suppose $N_1 = 4$. Then $N_2 \leqslant 7$. If $N_2 = 4$, then N_3 is arbitrary and so a possible vertex symbol is

$$(4, 4, m) \quad \text{for any } m \geqslant 4.$$

Now consider $(4, k, m)$, where $m \geqslant k > 4$. Figure 11 shows that both k and

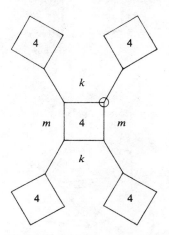

The vertex symbol $(4, k, m)$ says k and m are even.

Figure 11.

m must be even (since, for example, each k-gon is surrounded by 4-gons and m-gons in alternating order). Thus $k = 6$, and since $m \leqslant 11$ from (4), possible vertex symbols are

$$(4, 6, 6)$$
$$(4, 6, 8)$$
$$(4, 6, 10).$$

(c) Suppose $N_1 = 3$. We thus consider $(3, k, m)$. As in (a), $k = m$. And, as in (b), we see that k and m are even if they are greater than 3. Since $k \leqslant 11$ from (4), the possible vertex symbols are

$$(3, 3, 3)$$
$$(3, 4, 4)$$
$$(3, 6, 6)$$
$$(3, 8, 8)$$
$$(3, 10, 10).$$

All cases have now been exhausted and the list of possible vertex symbols is complete. Now we can arrange the resulting polyhedra in a reasonable order and give them their usual names.

(A) $(4, 4, m)$ for $m \geqslant 3$.

These are *prisms* with m-gons as bases.

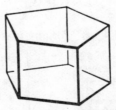

A polyhedron with vertex symbol $(4, 4, 5)$—a prism with 5-gons as bases.

Figure 12a.

(B) $(3, 3, 3, m)$ for $m \geqslant 3$.

These are *antiprisms* with m-gons as bases.

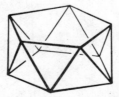

A polyhedron with vertex symbol $(3, 3, 3, 5)$—an antiprism with 5-gons as bases.

Figure 12b.

(C) A special case of (A) is $(4,4,4)$ *Cube.*
And a special case of (B) is $(3,3,3,3)$ *Octahedron.*
We also have $(3,3,3)$ *Tetrahedron,*
                                $(5,5,5)$ *Dodecahedron,*
                                $(3,3,3,3,3)$ *Icosahedron.*

These are the five regular polyhedra or *Platonic Solids* shown in Figure 3. (Note that each is composed of only one kind of *n*-gon.)

(D) The remaining thirteen semiregular polyhedra, called the *Archimedean Solids*, are shown below.

$(3,6,6)$ Truncated Tetrahedron

$(3,8,8)$ Truncated Cube

$(3,10,10)$ Truncated Dodecahedron

$(4,6,6)$ Truncated Octahedron

$(5,6,6)$ Truncated Icosahedron

(3, 4, 3, 4)   Cuboctahedron

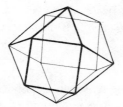

(3, 5, 3, 5)   Icosidodecahedron

(3, 4, 4, 4)   Rhombicuboctahedron

(4, 3, 4, 5)   Rhombicosidodecahedron

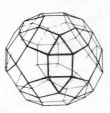

(4, 6, 8)   Great Rhombicuboctahedron

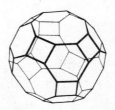

(4, 6, 10)   Great Rhombicosidodecahedron

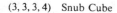 (3, 3, 3, 4) Snub Cube

 (3, 3, 3, 5) Snub Dodecahedron

The illustrations of these polyhedra have been reproduced from the book *Polyhedra Primer* [2] (copyright, 1978, by Peter Pearce and Susan Pearce) through the courtesy of the authors.

Readers interested in a more advanced treatment of this subject can consult [3].

REFERENCES

1. W. W. R. Ball and H. S. M. Coxeter, Mathematical Recreations and Essays, 12th ed., Univ. of Toronto Press, 1974.
2. P. Pearce and S. Pearce, Polyhedra Primer, Van Nostrand Reinhold, New York, 1978.
3. T. R. S. Walsh, Characterizing the vertex neighbourhoods of semiregular polyhedra, Geom. Dedicata, 1 (1972) 117–123.

Angles in Graphs

Frank Harary and David B. Leep, *University of Michigan, Ann Arbor, MI*

Graph theory is rapidly becoming a standard course in an undergraduate mathematics curriculum. In addition to its aesthetic and intuitive appeal, graph theory provides a wealth of elementary, yet useful, results which apply to other branches of mathematics—as, for example, topology, geometry, group theory, matrices, algebra, geometry, analysis, statistics, and number theory. Here is one dealing with trigonometry.

We define an angle determined by an ordered triple of distinct points in a connected graph using the conventional trigonometric law of cosines applied to the graphical distances. It is quickly seen that in several types of graphs, the only angles which occur are 0, $\pi/3$, and π. Graphs having such angles are called "angular." We shall show how to specify precisely which families of graphs are angular.

1. Preliminary Concepts. A *graph* G is a finite nonempty set V of points together with a set E of pairs of distinct points in V. An element $\{u, v\}$ in E is called the *line* joining u and v. Points u, v of V are called *adjacent* if there is a line joining them.

A *path* is an alternating sequence of distinct points and lines $v_0 e_1 \ldots v_{n-1} e_n v_n$ beginning and ending with points $v_0, v_n \in V$, where each e_i is the line $\{v_{i-1}, v_i\}$. A *cycle* is defined as a path that has $v_0 = v_n$ and $n \geqslant 3$. A path with n points is denoted P_n; a cycle with n points is written C_n.

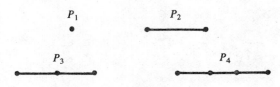

The four smallest paths.*

Figure 1.

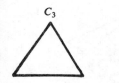

The three smallest cycles.*

Figure 2.

The *complete graph* K_n consists of n points such that every pair of distinct points are adjacent.

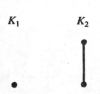

The five smallest complete graphs.

Figure 3.

We will also require the notion of complete multipartite graphs $K(r_1, r_2, \ldots, r_k)$. In these graphs with $n = \sum r_i$ points, the set V is partitioned into k disjoint sets $S_1, S_2, \ldots, S_k$ having $r_1, r_2, \ldots, r_k$ points, respectively such that two points are adjacent if and only if they belong to different sets S_i and S_j. We show such a graph in Figure 4b. A complete multipartite graph with $k = 2$ is called a *complete bipartite graph*, written $K(m, n)$ or $K_{m,n}$. Figure 5 exhibits three of these graphs. This illustrates an operation between two graphs that we need. The *join* of G_1 and G_2, denoted $G_1 + G_2$, is the graph whose points are those of G_1 and G_2, where two points u and v

---

*Observe that P_4' is a path and C_5' is a cycle.

Actually, P_4 and P_4' are isomorphic, and C_5 and C_5' are isomorphic, in the sense that there exists a 1-1 onto mapping between point sets which preserves adjacency. Graph theorists usually consider isomorphic graphs as being the same. Thus, P_4 denotes all 4-point path isomorphs, and C_5 denotes all 5-point cycle isomorphs.

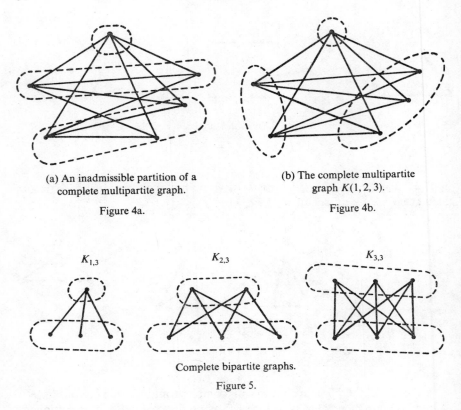

(a) An inadmissible partition of a   (b) The complete multipartite
   complete multipartite graph.   graph $K(1, 2, 3)$.

Figure 4a.   Figure 4b.

$K_{1,3}$ $K_{2,3}$ $K_{3,3}$

Complete bipartite graphs.

Figure 5.

are adjacent whenever u, v are already adjacent in G_1 or in G_2, or u, v are in opposite sets G_1, G_2. When $G_1 = K_2$ and G_2 consists of three isolated points, their join is displayed in Figure 6.

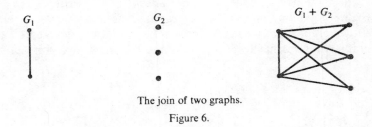

G_1 G_2 $G_1 + G_2$

The join of two graphs.

Figure 6.

A graph is *connected* if every pair of distinct points is joined by a path. Thus, complete graphs are connected. The converse, of course is false; the connected graph $G_1 + G_2$ (Figure 6) is not complete.

2. Distance and Angles in Graphs.

The *length* of a path is the number of lines it contains. A *geodesic* joining two points u, v is a shortest path joining u and v. The *distance* $d(u,v)$ is the length of any geodesic joining them. When $u = v$, we define $d(u,v) = 0$. The *diameter* $d(G)$ of a connected graph G is the length of any longest geodesic.

In a connected graph G, distance is a *metric*. That is, for all points u, v, w:

   (i) $d(u,v) \geqslant 0$, and $d(u,v) = 0$ if and only if $u = v$;

   (ii) $d(u,v) = d(v,u)$;

   (iii) the triangle inequality holds, $d(u,w) \leqslant d(u,v) + d(v,w)$.

Any three distinct points u, v, w in a connected graph G determine a *formal triangle* with sides of length

$$a = d(u,v), \quad b = d(v,w), \quad \text{and} \quad c = d(u,w),$$

as depicted in Figure 7.

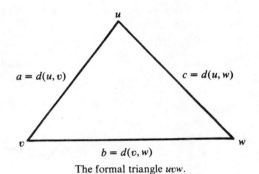

The formal triangle uvw.

Figure 7.

We now define the angle $\theta = \sphericalangle uvw$ by applying the Law of Cosines in this formal triangle:

$$\cos\theta = \frac{a^2 + b^2 - c^2}{2ab}.$$

Formal triangles may be rather unusual looking:

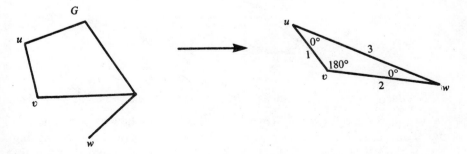

Example 1. Consider triangle uvw for $u, v, w \in G$.

The above notion of angle agrees with some of our experience:

(i)   A triple of pairwise equidistant points in a graph (an equilateral formal triangle) determines three angles of 60°.

(ii)   Base angles of an isosceles formal triangle are equal.

(iii) Any pythagorean triple of positive integers which appear as distances in a graph determines an angle of 90°.

(iv) $\cos\theta$ lies between -1 and $+1$.

(v)   The sum of the angles in a formal triangle is 180°. (This is seen by first noting that the sides of a formal triangle have positive integer lengths corresponding to a usual triangle in the plane, because of the triangle inequality. The angles of a triangle in the plane are given by the cosine law, and thus the angles in a formal triangle sum to 180°.)

The converse of (ii) is false. Example 1 demonstrates that equal angles of zero degrees in a formal triangle need not have equal opposite sides. One can easily verify (Figure 7) that

(vi) $\measuredangle v = \measuredangle w \Rightarrow$ either $a = c$ or $v = 0°$ or w.

It is worth noting that the "usual" additivity property of angular measurement does not hold.

Example 2.   The measure of $\measuredangle v$ in C_4 is 180°.

However,

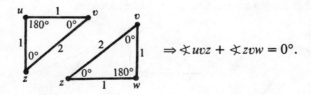

$\Rightarrow \measuredangle uvz + \measuredangle zvw = 0°.$

Similarly, $\measuredangle vuz = 180°$ in $K_{1,3}$, but $\measuredangle vuw + \measuredangle wuz = 360°$. (See p. 67.)

We shall henceforth only consider connected graphs with $n \geqslant 3$ points. By $\measuredangle uvw$ in such a graph is meant angle uvw in the formal triangle having u, v, w as vertices. Our objective is to find all connected graphs that contain only the angles 0°, 60°, 180°. We call such graphs *angular*. It is obvious

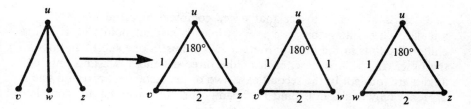

that if one angle of a formal triangle in an angular graph is 60°, then all three angles are 60° and [by property (vi)] the triangle is *equilateral*. And if one angle of a formal triangle is 0°, then the other two angles are 0° and 180°. A formal triangle is *degenerate* if the sum of two sides equals the third side. If we take $a + b = c$ in Figure 7, then the angles at u and w are 0°, and angle θ at v is 180°. This was illustrated in Examples 1 and 2. We collect these observations in the following statement.

Lemma. *A connected graph G is angular if and only if each formal triangle in G is equilateral or degenerate.*

Which graphs are angular? Some classes of angular graphs are described below.

1. Every angle of a connected graph G is 60° if and only if G is a complete graph K_n with $n \geqslant 3$ points. In one direction this is immediate as every formal triangle in K_n is equilateral. Conversely, if all angles are 60° in G, then every formal triangle must be equilateral. Since G has at least one line, all pairs are adjacent, and so G is K_n.

2. Every angle of a connected graph G is 0° or 180° if and only if G is either a path P_n (shown in Figure 1) or the 4-cycle C_4 (Figure 2). This can be shown by first observing that G contains no subgraph of the form C_3 or $K_{1,3}$. It follows that G is either a path P_n (as shown in Figure 1) with $n \geqslant 3$ or a cycle C_n with $n \geqslant 4$. But for $n \geqslant 5$, each cycle C_n contains a formal triangle with an angle different from 0° or 180°.

$$G = C_n \qquad (n \geqslant 5)$$

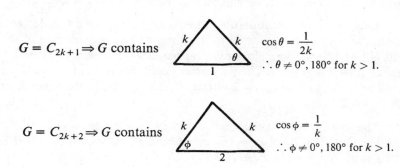

$G = C_{2k+1} \Rightarrow G$ contains $\qquad\cos\theta = \dfrac{1}{2k}$

$\therefore \theta \neq 0°, 180°$ for $k > 1$.

$G = C_{2k+2} \Rightarrow G$ contains $\qquad\cos\phi = \dfrac{1}{k}$

$\therefore \phi \neq 0°, 180°$ for $k > 1$.

3. It is also easy to see that every complete bipartite graph $K_{m,n}$ is angular. This follows because every triple of distinct points of $K_{m,n}$ has either all points in the same subset, or two points in one subset and one in the other. In the first case, each of the three distances is 2 and only 60° angles are formed. In the second case, two of the distances are 1 and one is 2. Hence only angles of 0° and 180° can occur. Note also that the complete bipartite graph $K_{2,2}$ is isomorphic to the 4-cycle C_4.

4. Two lines of a graph G are *independent* if they have no points in common. Now consider the graphs obtained from $K_{m,n}$ by removing a set S of independent lines (Figure 8). Note that $K_{3,3}$ less three independent lines is the cycle C_6.

 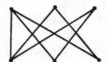

The graphs obtained on removing 1, 2, or 3 independent lines from $K_{3,3}$.

Figure 8.

Let us now verify that every graph obtained from $K_{m,n}$ ($m \geq 3, n \geq 2$), by removing a set of independent lines, is angular. To prove this, assume that S_1 and S_2 are the disjoint subsets of G that contain $m \geq 3$ and $n \geq 2$ points, respectively. Thus, the distance between any two points in the same S_i equals 2. Suppose $n \geq 3$. Then the removal of independent lines leaves the distance between any two points in different S_i equal to 1 or 3. Checking each formal triangle that remains shows that G is angular. If $n = 2$, at most two independent lines can be removed from $K_{m,2}$ to produce G. The argument above still applies when one line is removed. If two independent lines are removed, there are two points $v, v' \in S_1$ with $d(v, v') = 4$, whereas the distance between every other pair of points in the same S_i equals 2. Since the distance between two points in different S_i equals 1 or 3, it follows by checking the remaining formal triangles that G is angular.

5. Every complete multipartite graph is angular. To see this, let x, y, z be distinct points in $K(r_1, \ldots, r_k)$. The distance between any two of these points is 1 or 2. If the formal triangle xyz has sides of length 1, 1, 1 or 2, 1, 1 or 2, 2, 2, then it is angular. The only other possibility is 2, 2, 1, but this can't occur since $d(x, y) = d(x, z) = 2$ implies that x, y, z lie in the same S_i and so $d(y, z) = 2$.

6. The join of two angular graphs may or may not be angular.
(a) The join $P_3 + P_4$ of $P_3 = \{v_0, v_1, v_2\}$ and $P_4 = \{w_0, w_1, w_2, w_3\}$ is not angular since $\Delta w_0 w_2 w_3$ in $P_3 + P_4$ has sides of length 1, 2, 2.

(b) It follows, from the remarks in 1 and 5 above, that $K(r_1, r_2, \ldots, r_k) + K_n$ is angular for every complete multipartite graph $K(r_1, r_2, \ldots, r_k)$ and complete graph K_n. This is because $K(r_1, \ldots, r_k) + K_n$ is the same as $K(r_1, \ldots, r_k, 1, \ldots, 1)$, where $K_n = K(1, \ldots, 1)$.

3. Characterization of Angular Graphs.

The connected graphs illustrated in (1)–(6) of Section 2 are, in fact, the only examples of angular graphs. We shall now prove the following:

Theorem. *A connected graph G is angular if and only if G belongs to one of the following classes*:*

1. *A path P_n.*
2. *$K_{m,n} - S$ where $m \geqslant 3$, $n \geqslant 2$ and S is a (possibly empty) set of independent lines of G.*
3. *A complete multipartite graph $K(r_1, r_2, \ldots, r_k)$.*
4. *The join $K(r_1, r_2, \ldots, r_k) + K_m$ of a complete multipartite graph with a complete graph.*
5. *A complete graph K_n.*

Proof. We have already seen that each class (1)–(5) of connected graphs is angular. To prove the converse, we shall show that a connected angular graph must belong to one of the five classes above. The proof is divided into six parts by separately considering angular graphs of diameter $d = 1$ to 5 and $d \geqslant 6$. It is convenient to start the attack by finding the angular graphs of diameter 5, and then determine all those of higher diameter. The remaining cases will separately analyze the angular graphs of each smaller diameter.

We begin by assuming that G is angular and by noting (Lemma) that every formal triangle of G is either equilateral or degenerate.

Case 1. $d = 5$. There is a geodesic of length 5 in the angular graph G, and $d(u, v) \leqslant 5$ for every pair of points u and v. Let the distance $d(v_0, v_5) = 5$ and let $v_0 v_1 v_2 v_3 v_4 v_5$ be a geodesic joining them.

Let w_i be a point of G for which $d(v_0, w_i) = i$. There are two possibilities for $i = 5$: either a point $w_5 \neq v_5$ exists or it doesn't. If there is a $w_5 \neq v_5$, the formal triangle $\Delta v_0 w_5 v_5$ has $d(w_5, v_5) = 5$ since $d(v_0, v_5) = 5 = d(v_0, w_5)$, and this forces the third side $w_5 v_5$ to have length 5. (See figure on p. 70.)

---

*These classes are not mutually exclusive, e.g., $K_p = K(1, 1, \ldots, 1)$, and the removal of any two independent lines of $K_{2,3}$ yields the path P_5.

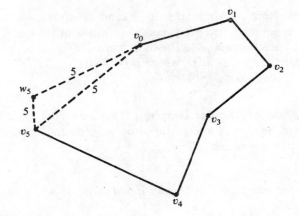

Similar results hold for $i = 3, 4$. Thus,

$$d(v_0, w_i) = i \Rightarrow d(w_i, v_i) = 1 \quad \text{for} \quad i = 3, 4, 5.$$

We summarize the above observations in the next sentence, where $\Delta v_0 w_i v_i$ denotes the existence of points v_0, w_i, v_i that form a formal triangle with vertices v_0, w_i, v_i.

(a) $\Delta v_0 w_i v_i \Rightarrow d(w_i, v_i) = i \quad \text{for} \quad i = 3, 4, 5.$

We now develop a sequence of similar observations. It can be seen that for $i = 4, 5$:

$$d(w_{i-1}, w_i) \neq 1 \Rightarrow \measuredangle v_0 \neq 0°, 60°, 180°.$$

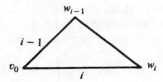

Therefore

$$d(v_0, w_{i-1}) = i - 1 \quad \text{and} \quad d(v_0, w_i) = i \Rightarrow d(w_{i-1}, w_i) = 1 \quad \text{for} \quad i = 4, 5$$

can be summarized as

(b) $\Delta v_0 w_{i-1} w_i \Rightarrow d(w_{i-1}, w_i) = 1 \quad \text{for} \quad i = 4, 5.$

Next, consider $\Delta v_{i-1} w_i v_i$ for $i = 4, 5$. From (a), we have $d(w_i, v_i) = i$. By definition, $d(v_{i-1}, v_i) = 1$. And since w_{i-1} can be taken as v_{i-1}, it follows from (b) that $d(v_{i-1}, w_i) = 1$. Therefore $\Delta v_{i-1} w_i v_i$ has sides of length $1, 1, i$ and this isn't angular when $i = 4$ or 5. This contradiction establishes that

(c) $\Delta v_{i-1} w_i v_i \Rightarrow w_i = v_i \quad \text{for} \quad i = 4, 5.$

A related argument can be used on $\Delta v_3 w_3 v_4$. Clearly $d(v_3, w_3) = 3$ follows from (a). Taking $w_3 = v_3$ and $w_4 = v_4$ in (b) yields $d(v_3, v_4) =$

$d(w_3, v_4) = 1$. Thus, $\Delta v_3 w_3 v_4$ is not angular, and this contradiction establishes

(d) $\Delta v_3 w_3 v_4 \Rightarrow w_3 = v_3$.

The next assertion considers $\Delta w_i v_0 v_5$ for $i = 1, 2$.

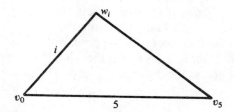

Since $\Delta w_i v_0 v_5$ is degenerate (it is not equilateral), $d(w_i, v_5) = 5 - i$ or $d(w_i, v_5) = 5 + i$. The latter is excluded since $d(G) = 5$. Therefore,

(e) $\Delta w_i v_0 v_5 \Rightarrow d(w_i, v_5) = 5 - i$ for $i = 1, 2$.

Our last concern is $\Delta w_i v_i v_5$ for $i = 1, 2$.

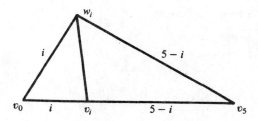

It follows from (e), that $d(w_i, v_5) = 5 - i$. Thus, $d(w_i, v_i) = 0$, $5 - i$, or $10 - 2i$. The choice $10 - 2i$ is excluded since $10 - 2i > 5 = d(G)$. And $d(w_i, v_i) = 5 - i$ is excluded since $\Delta v_0 v_i w_i$ would be neither equilateral nor degenerate. Therefore $d(w_i, v_i) = 0$. In other words,

(f) $\Delta w_i v_i v_5 \Rightarrow d(w_i, v_i) = 0$ for $i = 1, 2$.

We can now put this all together as follows: (c) shows that the only points at distance 4 and 5 from v_0 are v_4 and v_5, respectively; (d) shows that v_3 is the only point at distance 3 from v_0; and (f) establishes that the only points at distance 1 and 2 from v_0 are v_1 and v_2, respectively. Therefore, the graph of G must be the path P_6.

Case 2. $d > 5$.   Let G be an angular graph of diameter t, where $t \geqslant 6$. We now prove by induction on t that G is the path P_{t+1}. We take as the inductive hypothesis:

$G = P_{i+1}$ *for all angular graphs G with $d(G) = i$ $(5 \leqslant i < t)$.*

Let G be an angular graph with diameter $d(G) = t$. Pick a geodesic of length t, labeled $v_0 v_1 v_2 \ldots v_t$. Using the same reasoning as in Case 1 (a)–(c), we see there is no $w_{t-1} \neq v_{t-1}$ and no $w_t \neq v_t$. Let G' be the graph obtained from G by deleting v_t and the line $\{v_{t-1}, v_t\}$. This is the only line of G containing v_t. (If $\{w, v_t\}$ is a line of G, then $d(v_0, w) = t - 1$ or t, and so $w = v_{t-1}$.) Since any geodesic in G between two points distinct from v_t doesn't contain $\{v_{t-1}, v_t\}$, the distance between such points in G' is the same as their distance in G. Thus, G' is angular.

We now show that $d(G') = t - 1$. Suppose, to the contrary that $d(z_1, z_2) = t$ for $z_1, z_2 \in G'$. Then $z_1, z_2 \neq v_{t-1}$ (otherwise, $d(z_1, v_t)$ or $d(z_2, v_t)$ equals $t + 1$). Let j be defined as the distance $d(v_0, z_2)$. Then $j \neq t - 1$.

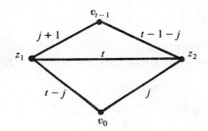

We now observe the following:

(a) $\Delta v_0 z_1 z_2 \Rightarrow d(v_0, z_1) = t - j.$

(b) $\Delta v_0 v_{t-1} z_2 \Rightarrow d(z_2, v_{t-1}) = t - 1 - j.$

(c) $\Delta z_1 z_2 v_{t-1} \Rightarrow d(z_1, v_{t-1}) = j + 1.$

(d) $\Delta v_0 z_1 v_{t-1} \Rightarrow d(z_1, v_{t-1}) = \begin{cases} j - 1 \text{ if } t - j \neq t - 1, \\ t - 1 \text{ if } j = 1. \end{cases}$

If $j \neq 1$, then (c) and (d) give a contradiction. If $j = 1$, then (c) and (d) give $2 = j + 1 = t - 1$ which is a contradiction since $t \geq 6$. This establishes that $d(G') = t - 1$.

By the inductive hypothesis, $G' = P_t$. Therefore, $G = P_{t+1}$ since $G = P_t \cup \{v_t\} \cup \{v_{t-1}, v_t\}$.

Together, cases 1 and 2 show that $G = P_{t+1}$ for an angular graph G with diameter $d(G), t \geq 5$.

Case 3. $d = 4$. Assume that $v_0 v_1 v_2 v_3 v_4$ is a geodesic in the angular graph G. The same methods used earlier can be used to show that:

(a) There are no points $w_3 \neq v_3$ and $w_4 \neq v_4$.

(b) There is no point $w_1 \neq v_1$.

(c) $(w_2, v_1) = 1 = d(w_2, v_3)$.

(d) $d(w_2, v_4) = 2$.

(e) If there are two points w_2, w_2' at distance 2 from v_0, then $d(w_2, w_2') = 2$.

Conditions (a)–(e) determine the adjacencies of all the points of G. The graph of G, satisfying (a)–(e), is shown below. If there are no points $w_2 \neq v_2$, then $G = P_5$. Otherwise, G is obtained from $K_{r,2}$ (with $r \geqslant 3$) by removing two independent lines as shown in Figure 9.

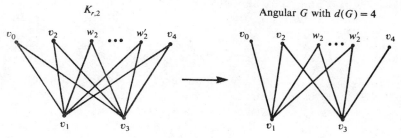

Figure 9.

Case 4. $d = 3$. Let w_i, w_i' denote points that are distance i from v_0. Using the same strategy of considering formal triangles, and the distances between pairs of points which are implied by these triangles, one can show that:

(a) There is no point $w_3 \neq v_3$.
(b) $d(w_2, v_3) = 1$, $d(w_1, v_3) = 2$.
(c) $d(w_2, w_2') = 2$, $d(w_1, w_1') = 2$.
(d) $d(w_1, w_2) = 1$ or 3.
(e) If $d(w_1, w_2) = 3$, then $d(w_1', w_2) = 1$ for all $w_1' \neq w_1$.
(f) If $d(w_1, w_2) = 3$, then $d(w_1, w_2') = 1$ for all $w_2' \neq w_2$.

An angular graph satisfying conditions (a)–(f) is shown below.

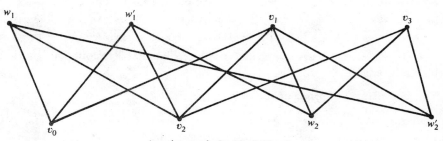

Angular graph G with $d(G) = 3$.

Observe that an angular graph with diameter $d(G) = 3$ is either obtained by removing any one line from $K_{r,2}$ with $r \geqslant 2$ (as when one or both w_1's and w_2's are not present) or is constructed by removing a nonempty set S of independent lines from $K_{r,s}$ with both $r, s \geqslant 3$ (as when both w_1's and w_2's are present).

The last two cases show that every angular graph of diameter 3 or 4 falls into class 2 of the theorem.

Case 5. $d = 2$. Suppose G is angular and $d(G) = 2$. Then there is a point v_0 that is not adjacent to some other point. Let $S_1 = \{v \in V : v \text{ is not adjacent to } v_0\}$, and let S'_2 be the complement $S'_2 = V - S_1$. Let G'_2 be the subgraph induced by the point set S'_2, and let G_1 be induced by S_1. We make two key assertions.

Claim 1. G_1 is totally disconnected (i.e., has no lines).

Let $v \in S_1$ ($v \neq v_0$). Then v is not adjacent to v_0 by the definition of S_1. Let $v, v' \in S_1$ be distinct from v_0. Since $d(G) = 2$ and v, v' are not adjacent to v_0, we have $d(v_0, v) = d(v_0, v') = 2$. Since $\Delta v_0 v v'$ is a formal triangle in the angular graph G_1, it follows that $d(v, v') = 2$. Therefore, v, v' are not adjacent. Thus, any two points in S_1 are not adjacent, and so G_1 is totally disconnected.

Claim 2. $G = G_1 + G'_2$ (i.e., G is the join of G_1 and G'_2).

We must show that each point $y \in S'_2$ is adjacent to each point $x \in S_1$. If $x = v_0$, then x and y are adjacent by the definition of S'_2. If $x \neq v_0$, then $d(v_0, x) = 2$ (by Claim 1) and $d(v_0, y) = 1$. Since G is angular, $\Delta v_0 x y$ cannot have sides of length $1, 2, 2$. Therefore, $d(x, y) = 1$. Thus, $G = G_1 + G_2$.

If $d(G'_2) = 2$, apply the previous step to construct $G'_2 = G_2 + G'_3$, where G_2 is totally disconnected. Continue this process until $G'_{n-1} = G_{n-1} + G'_n$, where either $d(G'_n) = 1$ or G'_n has no lines. If $d(G'_n) = 1$, then G'_n is complete. Since $d(G) = 2$, either $G = G_1 + \cdots + G_n$ where each G_i is totally disconnected, or $G = G_1 + \cdots + G_{n-1} + G_n$ where G_n is complete and every other G_i is totally disconnected. These are the graphs in classes 3 and 4 of the theorem. Writing r_i for the number of points in G_i, we have $G = K(r_1, \ldots, r_n)$ or $G = K(r_1, \ldots, r_{n-1}) + K_{r_n}$.

Case 6. $d = 1$. Every connected graph of diameter 1 is complete and is therefore angular.

This completes the proof of the theorem.

REFERENCE
1.   F. Harary, *Graph Theory*, Addison-Wesley, Reading, MA, 1969.

3. NUMBER THEORY

The theory of numbers, as a distinct branch of mathematics, can be said to have originated with Fermat (1601–1665). Many beautiful and far-reaching theorems about prime numbers, the building blocks of number theory, have been established since then. And yet, there still exist a large number of intriguing questions and enticing conjectures that remain unresolved to date. This, to a large extent, reflects the magic charm of number theory: mathematicians and amateurs alike continue to be captivated by the easily comprehended, yet challenging, nature of the subject.

Some of the most absorbing pursuits in number theory concern the nature and dispersion of the primes. For instance, consider the question: *Where, and with what density, do the primes reside among the positive integers?* Partial answers are available. In 1896, Hadamard and Vallée-Poussin independently proved that the number of primes not exceeding n approaches $n/(\ln n)$ for large n. This key result, known as the Prime Number Theorem, provides accurate information about how many primes are less than or equal to a given natural number (and how many primes "tend" to lie between two natural numbers), but it does not tell us which primes these are. The complete answer to our question still clamors for attention. It raises, and is interwoven with, many other related questions, such as: *Which numbers of a specified form (say, $2p + 1$ or $2^p - 1$ for prime p) are prime, and how many such primes are there?*

The first two articles of this chapter illustrate some of the above concerns. Dirichlet (1805–1859) proved that there is an infinite number of primes of the form $\alpha n + \beta$, when α and β are relatively prime. Thus, the set of primes $\{10n + 1\}$ ending in 1 is infinite. In his article "Initial Primes," R. L. Francis considers and sharpens the analagous result for primes whose leading digit is 1 (as, for example, 17 and 127). His paper also discloses the minimum number of such primes that can be found between $n!$ and $(n + 1)!$ for each natural number n.

The greatest known primes, called Mersenne primes, are of the form $2^p - 1$ (p prime). Marin Mersenne's original list $\{M_p : p = 2, 3, 5, 7, 13, 19, 31, 67, 127, 257\}$, sent to Fermat in 1664, contained several errors. It omitted M_{61} (proved to be prime around 1880), and it included M_{67} (proved composite in 1903) and M_{257} (found to be composite in 1931). Evidently progress in the study of the Mersenne primes was slow. The 15th Mersenne prime M_{1279} was discovered in 1953; the 23rd prime $M_{11,213}$ was found in 1963. It took an IBM computer 40 minutes, in 1971, to determine that $M_{19,937}$ was the 24th Mersenne prime. The 25th prime $M_{21,701}$ (containing 6,533 digits) was found in 1978. Finally, the 27th, and greatest known, prime $M_{44,479}$ (containing 13,395 digits) was discovered in 1979. The numbers involved have become truly enormous! How, and by what means, were these primes discovered? In "Mersenne Primes and the Lucas-Lehmer Test," W. G. Leavitt describes one special technique that was used for testing the primality of M_p. Leavitt's article will reaffirm the belief that major inroads into the mysteries of Mersenne primes (viz: are there infinitely many?) will come from man's head-held calculator assisted by modern computers, and not vice versa.

For thousands of years, people have been irresistibly fascinated by properties of the integers. The ancient Chinese and Babylonians, for example, came under this spell more than 4,000 years ago. Diophantus (circa 275) also succumbed to this spell, and his distinctive work, focusing on integral solutions to algebraic equations, seduced others into the fold. Fermat was a victim. A translation of Diophantus's *Arithmetica* excited him and somewhere, in the margin of his copy, lies the most celebrated, and still unsolved, of all Diophantine problems.

Interest in the study of the integers continues unabated today. A good illustration is the article " An Introduction to the Tarry-Escott Problem" by Harold Dorwart and Warren Page. The problem considered here is that of finding two sets of integers $\{\alpha_1, \alpha_2, \ldots, \alpha_s\}$ and $\{\beta_1, \beta_2, \ldots, \beta_s\}$ such that

$$\sum_{i=1}^{s} \alpha_i^j = \sum_{i=1}^{s} \beta_i^j \quad \text{for each } j = 1, 2, \ldots, k.$$

The authors show how to generate solutions to this problem, and they consider some interesting properties of these solutions.

The digital root of a natural number is the digit obtained by successively adding the digits until a single digit is reached. Thus, $DR(25) = 7$ and $DR(92) = 2$. Thomas Dence's article "The Digital Root Function" is a fitting way to conclude this chapter since it exemplifies the "easy conjecture —challenging proof" nature of number theory. Dence's study, using a computer to gather data, serves as an excellent vehicle for students to appreciate the interplay of number-theoretic concepts with other branches of mathematics.

Initial Primes

Richard L. Francis, *Southeast Missouri State University, Cape Girardeau, MO*

Any prime number, other than 2 or 5, must have a terminal digit of 1, 3, 7, or 9. Therefore, by Dirichlet's theorem,

> The set of primes of the form $an + b$ is infinite if and only if the integers a and b are relatively prime,

it is seen that each set of primes ending in one of the above four digits is infinite. For example, primes ending in 1 are of the form $10n + 1$. Since 10 and 1 are relatively prime, the set of primes of the form $10n + 1$ is infinite.

This note focuses on the analogous result by considering the leading digit of a prime. For our purposes, a prime number will be called an *initial prime* if its leading digit is the number 1 (e.g., 13 or 127).

According to Bertrand's Conjecture (established in 1850 by Tchebycheff):

> For each natural number $n > 1$, there exists at least one prime between n and $2n$.

Now consider the integers between 10^{n-1} and $2 \cdot 10^{n-1}$. All such integers interior to this interval have an initial digit of 1. Therefore, there exists at least one initial prime in this interval. Since 10^{n-1} and $2 \cdot 10^{n-1}$ each contain n digits, any initial prime in this interval also has n digits in its representation. Therefore, *for each natural number n, there is an initial prime containing exactly n digits*. And it is also clear that *the set of initial primes is infinite*.

Since $n! < 2(n!) \leq (n + 1)!$ for $n \geq 1$, we see (by Bertrand's Conjecture) that there exists a prime between any two consecutive factorials. In fact, we can use this to show that:

> **If $n \geq 2^m - 1$, there exist at least m primes between $n!$ and $(n + 1)!$.**

This is clear since $n! < 2 \cdot n! < 2^2 \cdot n! < 2^3 \cdot n! < \cdots < 2^{m-1} \cdot n! < 2^m \cdot n! \leq (n + 1)!$.

Are any of these m primes between $n!$ and $(n + 1)!$ initial primes? The answer is affirmative when n is sufficiently large. Specifically:

If $n \geqslant 10^k - 1$ $(k \geqslant 1)$, there exist at least $k - 1$ initial primes between $n!$ and $(n + 1)!$.

For verification, assume that $n!$ is expressed in scientific notation as $n! = d \cdot d_1 d_2 \ldots \times 10^j$, where $d, d_1, d_2, \ldots$ are digits and $d \neq 0$. Since $1 \leqslant d \leqslant 9$, it is clear that

$$n! < 10 \times 10^j < 10^2 \times 10^j < 10^3 \times 10^j < \cdots$$
$$< 10^{k-1} \times 10^j < 10^k \times 10^j < 10^k \cdot n! \leqslant (n + 1)!.$$

Moreover,

$$2 \cdot 10^i < 10^{i+1} \qquad \text{for each} \quad i = 1, 2, \ldots, k - 1,$$

and so there exists at least one initial prime between 10^i and 10^{i+1} for each $i = 1, 2, \ldots, k - 1$.

Remark. There will exist an initial prime between $10^k \times 10^j$ and $10^k \cdot n!$ if $2 \leqslant d \cdot d_1 d_2 \ldots$.

Example

(a) There exist at least 6 primes between $99!$ and $100!$ since $n = 99 > 2^m - 1$ for $m = 6$.

(b) There is at least one initial prime between $99!$ and $100!$ since $n = 99 = 10^2 - 1$. Similarly, there are at least five initial primes between $(10^6 - 1)!$ and $10^6!$.

Mersenne Primes and the Lucas-Lehmer Test

W. G. Leavitt, *University of Nebraska, Lincoln, NE*

The largest known prime numbers are all of the form $M_p = 2^p - 1$, and they have come to be called "Mersenne primes" after Marin Mersenne (1588–1648), who in 1644 sent a list of such primes to Pierre Fermat (1601–1665). This original list, in fact, contained errors both of omission and of commission (see [1, p. 14]).

Clearly M_p is prime when $p = 2, 3, 5,$ and 7. In fact, one can easily see that when M_p is prime, p itself must be prime. The converse however is false as evidenced by $M_{11} = (23)(89)$. The list of those values of p for which M_p is prime (given at the end) falls into no pattern. Although new Mersenne primes continue to be found, it isn't even known if there are an infinite number of them.* There are now only 27 known Mersenne primes, and each newly discovered one makes quite a splash (as the "world-record largest prime"). In 1963, for example, the Iliac computer of the University of Illinois discovered $M_{11,213}$ to be prime, and for quite a while letters from that university carried the postmark "$2^{11,213} - 1$ is prime."

A description of the work (testing M_p for all primes $p < 21,000$) leading to the discovery of the 24th Mersenne prime $M_{19,937}$ is given in [2]. This number has 6,002 digits. The 25th Mersenne prime $M_{21,701}$ (having 6,533 digits) was discovered in October 1978 by two 18-year-old students in California. The current record is $M_{44,497}$ discovered by Harry Nelson and David Slowinski in May 1979.

One might wonder how it is known that $M_{44,497}$ is prime. This is truly an enormous number: A quick computation shows that it has 13,395 digits, so that simply to write it out would take several pages. Thus ordinary methods (such as testing whether or not it is divisible by any odd number less than its square root), even with the use of a giant computer, are completely out of the question. It happens, however, that there are special tests for the primality of the numbers M_p. The first such test, discovered by E. Lucas

---

*A brief history of Mersenne numbers and a complete calculation of all the digits of M_{521}, M_{607}, and M_{1279} is given by Horace S. Uhler [3]. Also see [4] and [6].

(1842–1891), was subsequently improved and simplified by D. H. Lehmer (1905–   ). The tests are extremely simple, and it is surprising that so few people know any of them. In what is to follow, we will discuss one such "Lucas-Lehmer" test.

Let $s_1 = 4$ and define $s_{n+1} = s_n^2 - 2$ for all $n \geqslant 1$. Clearly $M_2 \nmid s_1$ since 3 does not divide 4. Note, however, that $M_3 | s_2$ and $M_5 | s_4$. But $M_{11} \nmid s_{10}$. All this is reflected in the following Lucas-Lehmer test:

Theorem. M_p ($p > 2$) is prime if and only if $M_p | s_{p-1}$.

Note, that in testing for the primality of M_p, it is not necessary to calculate the rapidly increasing sequence $s_1, s_2, \ldots, s_{p-1}$. In fact, if we instructed our computer to start with $s_1 = 4$ and use the rule $s_{n+1} = s_n^2 - 2$, we would soon be beyond the capacity of even the largest computer (or indeed the capacity of the whole known universe). Thus we add the instruction: If $s_m > M_p$, divide by M_p and let the remainder be denoted by $\bar{s}_m$. Then proceed by using $\bar{s}_{n+1} = \bar{s}_n^2 - 2$ for all $n \geqslant m$. Clearly, M_p divides s_m if and only if it divides $\bar{s}_m$. And, this holds for all $n \geqslant m$. For example, one could test $M_5 = 31$ by calculating $s_1 = 4$, $s_2 = 14$, $s_3 = 194$; so $\bar{s}_3 = 8$ and $\bar{s}_4 = \bar{s}_3^2 - 2 = 62$, which is divisible by 31.

To see why the Lucas-Lehmer test works, we recall some well-known elementary facts:

(I)   If p is prime, the integers $\{n \not\equiv 0 \pmod{p}\}$ form a group relative to multiplication (mod p);

(II)   (Fermat) $a^{p-1} \equiv 1 \pmod{p}$ if p is prime and $p \nmid a$.

We use these to prove the following results.

Lemma 1. *If q is a prime such that $q = 8k + 7$ for some $k \geqslant 0$, then* $2^{(q-1)/2} \equiv 1 \pmod{q}$.

Proof. Multiply the even numbers from 2 to $q - 1$ to get

$$2 \cdot 4 \cdots (8k + 6) = 2^{4k+3}(4k + 3)!. \tag{1}$$

We also have congruences

$$4k + 4 \equiv -(4k + 3) \pmod{q}$$

$$4k + 6 \equiv -(4k + 1) \pmod{q}$$

$$\vdots \qquad \vdots$$

$$8k + 6 \equiv -1 \pmod{q}.$$

There are $2k + 2$ of these congruences, so we get

$$[2 \cdot 4 \cdots (4k + 2)][(4k + 4) \cdots (8k + 6)] \equiv (-1)^{2k+2}(4k + 3)! \pmod{q}.$$

$$\tag{2}$$

Since $4k + 3 < q$, it is clear that $q \nmid (4k + 3)!$ and so $(4k + 3)! \not\equiv 0(\text{mod } q)$ has an inverse mod q. Thus if we equate (1) and (2) and multiply by this inverse, we obtain the desired result

$$2^{(q-1)/2} = 2^{4k+3} \equiv 1(\text{mod } q).$$

Lemma 2. *If q is a prime such that $q = 12k + 7$ for some $k \geqslant 0$, then $3^{(q-1)/2} \equiv -1(\text{mod } q)$.*

Proof. Take the product of multiples of 3, from 3 to $18k + 9$, to get

$$3 \cdot 6 \cdots (18k + 9) = 3^{6k+3}(6k + 3)!. \tag{3}$$

We can partition these $6k + 3$ integers $\{3, 6, \ldots, 18k + 9\}$ into three groups: The first group consists of the $2k + 1$ integers $\{3, 6, \ldots, 6k + 3\}$; the second group consists of the next $2k + 1$ members written as

$$6k + 6 \equiv -(6k + 1)(\text{mod } q)$$
$$6k + 9 \equiv -(6k - 2)(\text{mod } q)$$
$$\vdots \qquad \vdots$$
$$12k + 6 \equiv -1(\text{mod } q);$$

and the third group contains the remaining $2k + 1$ integers written as

$$12k + 9 \equiv 2(\text{mod } q)$$
$$12k + 12 \equiv 5(\text{mod } q)$$
$$\vdots \qquad \vdots$$
$$18k + 9 \equiv (6k + 2)(\text{mod } q).$$

Taking the product of these three groups, we get

$$3 \cdot 6 \cdot 9 \cdots (18k + 9) \equiv (-1)^{2k+1}(6k + 3)! (\text{mod } q). \tag{4}$$

Since $6k + 3 < q$, we have $(6k + 3)! \not\equiv 0(\text{mod } q)$. Thus, as in Lemma 1, equating (3) and (4) yields

$$3^{(q-1)/2} = 3^{6k+3} \equiv -1(\text{mod } q).$$

We now proceed to a proof of the theorem. Our proof, essentially that of W. Sierpinski ([5, pp. 336–340]), will have several numbered formulas, each of which may be checked by some simple (and obvious) algebra. We invite the reader to do this checking.

We begin by defining some auxiliary quantities

$$u_n = \frac{a^n - b^n}{a - b} \quad \text{and} \quad v_n = a^n + b^n,$$

where $a = 1 + \sqrt{3}$ and $b = 1 - \sqrt{3}$. For odd n, the binomial theorem yields

$$v_n = 2\left(1 + \binom{n}{2} \cdot 3 + \binom{n}{4} \cdot 3^2 + \cdots + \binom{n}{n-1} \cdot 3^{(n-1)/2}\right) \qquad (5)$$

and

$$u_n = \binom{n}{1} + \binom{n}{3} \cdot 3 + \cdots + \binom{n}{n-2} \cdot 3^{(n-3)/2} + 3^{(n-1)/2}. \qquad (6)$$

And there are similar formulas when n is even. Thus u_n, v_n are integers for all $n \geq 1$.

The first formulas to check are

$$v_n^2 = v_{2n} + 2(-2)^n \quad \text{and} \quad u_{2n} = u_n v_n. \qquad (7)$$

Next observe that $2s_1 = 8 = v_2$, which is the first step in a simple induction proof that

$$(2)^{2^{n-1}} s_n = v_{2^n} \qquad \text{for all} \quad n = 1, 2, \ldots . \qquad (8)$$

Sufficiency Proof. M_p *is prime if* $M_p \mid s_{p-1}$.

It will be assumed, in this sufficiency proof, that q is a prime divisor of M_p. (Note that q is odd since M_p is an odd number.) Our objective, of course, is to show that $q = M_p$ and thereby conclude that M_p is prime. We proceed as follows.

Fact 1. $q \mid v_{2^{p-1}}$ *and* $a \mid u_{2^p}$.

The first part follows from (8) since $q \mid M_p$ and $M_p \mid s_{p-1}$. The second half of Fact 1 follows from the second part of (7).

We now define the set

$$S_q = \{n \geq 1 : q \mid u_n\}.$$

Since S_q is nonempty (it contains 2^p), there exists a least member $d \in S_q$. Now check

$$2u_{m+n} = u_m v_n + u_n v_m. \qquad (9)$$

And when $m > n$,

$$(-2)^{n+1} u_{m-n} = u_n v_m - u_m v_n. \qquad (10)$$

If $m, n \in S_q$, then (9) shows that $m + n \in S_q$, and (for $m > n$) formula (10) shows that $m - n \in S_q$. These two properties of S_q lead to

Fact 2. *If* $m \in S_q$, *then* $d \mid m$.

If $d \nmid m$ for $m \in S_q$, then $m = jd + r$ for some integers j and r ($0 < r < d$). Since $d \in S_q$, it follows (by the first property) that $kd \in S_q$ for all integers $k \geq 1$. In particular, $jd \in S_q$ and (by the second property) $r = m -$

$jd \in S_q$. This however contradicts the definition of d. Therefore Fact 2 holds.

Since (Fact 1) $2^p \in S_q$, we know (Fact 2) that $d \mid 2^p$. Actually,

Fact 3. $d = 2^p$.

Suppose not. Then $d \mid 2^{p-1}$ since d is a power of 2. Thus $2^{p-1} = jd$ for some integer $j \geqslant 1$ and (as we argued above) then $2^{p-1} = jd \in S_q$. This means that $q \mid u_{2^{p-1}}$. At the same time (Fact 1), $q \mid v_{2^{p-1}}$. Therefore the formula (check it)

$$(v_n)^2 - 12(u_n)^2 = (-2)^{n+2} \tag{11}$$

requires that q divide a power of 2, which it can't do. This contradiction establishes Fact 3.

Since $\binom{q}{a} = q!/a!(q-a)!$ and q is prime, we see that $a \mid \binom{q}{a}$ for $a < q$. Therefore (5) and (6) yield

$$q \mid v_q - 2 \quad \text{and} \quad q \mid u_q - 3^{(q-1)/2}. \tag{12}$$

Our objective is to establish

Fact 4. $q \mid u_{q+1}$.

Write the second part of (12) as

$$u_q \equiv 3^{(q-1)/2} (\text{mod } q),$$

which by squaring becomes

$$u_q^2 \equiv 3^{q-1} (\text{mod } q).$$

Since Fermat's theorem (II) gives $3^{q-1} \equiv 1 \ (\text{mod } q)$, we see that

$$u_q^2 \equiv 1 \ (\text{mod } q). \tag{13}$$

Similarly, the first part of (12) yields $v_q \equiv 2 \ (\text{mod } q)$. Squaring gives

$$v_q^2 \equiv 4 \ (\text{mod } q). \tag{14}$$

Then (13) and (14) combined show that $4u_q^2 - v_q^2 \equiv 0 \ (\text{mod } q)$, that is,

$$q \mid 4u_q^2 - v_q^2.$$

Since $u_1 = 1$ and $v_1 = 2$, formulas (9) and (10) yield

$$2u_{q+1} = 2u_q + v_q \tag{15}$$

$$4u_{q-1} = 2u_q - v_q. \tag{16}$$

Thus

$$8u_{q+1}u_{q-1} = 4u_q^2 - v_q^2$$

and since q divides the right-hand side, either $q \mid u_{q-1}$ or $q \mid u_{q+1}$.

If $q \mid u_{q-1}$, then $q - 1 \in S_q$, and so $d \leqslant q - 1$. By our original assumption, $q \mid M_p$. Therefore $q \leqslant 2^p - 1$. But (Fact 3) $d = 2^p$. The ensuing contradiction $2^p \leqslant q - 1 \leqslant 2^p - 2$ shows that $q \nmid u_{q-1}$. Therefore $q \mid u_{q+1}$ as asserted.

Our final fact completes the sufficiency proof.

Fact 5. $q = M_p$, and so M_p is prime.

Since (Fact 4) $q \mid u_{q+1}$, we have $q + 1 \in S_q$ and $2^p = d \leqslant q + 1$. But, as above, $q \leqslant 2^p - 1$. Thus $2^p \leqslant q + 1 \leqslant 2^p$. Therefore $q = 2^p - 1 = M_p$.

Necessity Proof: M_p *is prime only if* $M_p \mid s_{p-1}$.

Assuming that M_p (where $p \geqslant 3$) is prime, we aim to show that $M_p \mid s_{p-1}$. Let $q = M_p$ in this proof. Then

Fact 6. $q = 24h + 7$ *for some integer* $h \geqslant 0$.

We know that p is odd, so $p - 1 = 2s$ for some s. Thus $q - 1 = 2^p - 2$ $= 2(2^{p-1} - 1) = 2(4^s - 1) = 2(4 - 1)(4^{s-1} + \cdots) = 6t$ for some integer t. Since $p \geqslant 3$, we also have $8 \mid 2^p$. Thus $8 \mid q + 1$ and $q - 1 = 8k + 6$ for some integer $k \geqslant 0$. Equating expressions for $q - 1$ gives

$$6t = 8k + 6.$$

Therefore $3 \mid k$ and $k = 3h$ for some $h \geqslant 0$. In particular, $q = 24h + 7$ and Fact 6 holds.

We now move on to

Fact 7. $v_{2^p} - (v_{2^{p-1}})^2 + 4 \equiv 0 \pmod{q}$.
First rewrite (7) as

$$v_{2n} - (v_n)^2 = (-2)^{n+1},$$

and substitute $n = 2^{p-1}$ to get

$$v_{2^p} - (v_{2^{p-1}})^2 = (-4) \cdot (2)^{2^{p-1}-1}. \tag{17}$$

From Lemma 1, and the fact that $2^p - 1 = q$, we have

$$(2)^{2^{p-1}-1} = 2^{(q-1)/2} \equiv 1 \pmod{q}.$$

Substitution into (17) establishes Fact 7.

Fact 8. $(v_{2^{p-1}})^2 \equiv 0 \pmod{q}$.

First check that

$$2v_{m+n} = v_m v_n + 12u_m u_n. \tag{18}$$

When specialized to $m = q$ and $n = 1$ (remember $q + 1 = 2^p$), this becomes

$$2v_{2^p} = v_q v_1 + 12u_q u_1 = 2v_q + 12u_q,$$

which can be rewritten

$$v_{2^p} = (v_q - 2) + 6(u_q + 1) - 4. \tag{19}$$

Now (12) gives $u_q \equiv 3^{(q-1)/2} \pmod{q}$. Fact 6 and Lemma 2 give $3^{(q-1)/2} \equiv -1 \pmod{q}$. Therefore, $u_q \equiv -1 \pmod{q}$ and $u_q + 1 \equiv 0 \pmod{q}$. Since $v_q \equiv 2 \pmod{q}$ by (12), we end up with

$$v_{2^p} \equiv -4 \pmod{q}.$$

When substituted into Fact 7, this yields exactly Fact 8.

Finally, it is now an easy matter to establish

Fact 9. $M_p \mid s_{p-1}$.

Fact 8 shows that the prime q satisfies $q \mid v_{2^{p-1}}$. Taking $n = p - 1$ in (8), we get $q \mid s_{p-1}$. But $q = M_p$, and so Fact 9 is proved.

The following is the list of all p for which M_p is known to be prime.

2	17	107	2203	9689	23209
3	19	127	2281	9941	44497
5	31	521	3217	11213	
7	61	607	4253	19937	
13	89	1279	4423	21701	

We conclude this discussion of Mersenne primes with a few points of interest.

The ancient Greeks called a number "perfect" if it equals the sum of its divisors (not including itself). Thus $28 = 1 + 2 + 4 + 7 + 14$ is perfect, and $28 = 2^2 M_3$. Euclid was aware of the fact that $2^{p-1} M_p$ is perfect when M_p is prime. It is a good exercise to show that these are the only even perfect numbers. (It is not known if there are odd perfect numbers.) Thus $2^{44,496} M_{44,497}$ is a perfect number (with 26,790 digits).

Another interesting fact about Mersenne primes is the following:

> *The sum of all the divisors of an integer $n \geqslant 2$ is a power of 2 if and only if n is a product of distinct Mersenne primes.*

For example, $n = 217 = (31)(7) = (M_5)(M_3)$ and $\sum_{d \mid 217} d = 256 = 2^8$. (This is problem E2493 in the October 1974 issue of the *American Mathematical Monthly*.)

REFERENCES

1. A. H. Beiler, Recreations in the Theory of Numbers—the Queen of Mathematics Entertains, Dover, New York, 1964.
2. Bryant Tuckerman, The 24th Mersenne prime, Proc. Nat. Acad. Sci. USA, 68, 10 (Oct. 1971) 2319–2320.
3. Horace S. Uhler, A brief history of the investigation on Mersenne numbers and the latest immense primes, Scripta Math., 18 (1952) 122–131.
4. ———, Full values of the first seventeen perfect numbers, Scripta Math., 20(1954) 240.
5. Waclaw Sierpinski, Elementary theory of numbers, Vol. 42, Mon. Mat. Polish Akad. Sci. (transl. by A. Hulanicki), 1963.
6. Donald B. Gillies, Three new Mersenne primes and statistical theory, Math. Comp., 18 (1964) 93–97.

An Introduction to the Tarry-Escott Problem

Harold L. Dorwart, *Glastonbury, CT,* and Warren Page, *New York City Technical College, Brooklyn, NY*

The Tarry-Escott Problem can be stated as a problem in elementary algebra or as a problem in number theory. E. B. Escott [9] wanted to find two polynomials of the same degree, each having integral roots and having the same first $k + 1$ coefficients. In other words, Escott sought polynomials, with integral roots, satisfying:

$$p(x) = \sum_{i=0}^{s} a_i x^{s-i} \quad \text{and} \quad q(x) = \sum_{i=0}^{s} b_i x^{s-i}$$

$$\text{with} \quad a_i = b_i \quad \text{for} \quad 0 \leqslant i \leqslant k. \tag{E}$$

G. Tarry [11], on the other hand, was concerned with finding sets of integers $\{\alpha_1, \ldots, \alpha_s\}$ and $\{\beta_1, \ldots, \beta_s\}$ that satisfy:

$$\sum_{i=1}^{s} \alpha_i^j = \sum_{i=1}^{s} \beta_i^j \quad \text{for each} \quad j = 1, 2, \ldots, k. \tag{T}$$

It is not too difficult to show* that the sets of roots $\{\alpha_1, \ldots, \alpha_s\}$ of $p(x)$, and $\{\beta_1, \ldots, \beta_s\}$ of $q(x)$, satisfy (T). Conversely, each solution set of (T) can be used to generate polynomials, $p(x) = \prod_{i=1}^{s}(x - \alpha_i)$ and $q(x) = \prod_{i=1}^{s}(x - \beta_i)$, that can be written as in (E). Thus, the two problems are equivalent.

Tarry's formulation of the problem quickly yields some simple procedures for finding solutions to particular cases, and so we shall use his approach. It will be convenient to call a solution of (T) a kth *degree solution*, written

$$\{\alpha_1, \ldots, \alpha_s\} \stackrel{k}{=} \{\beta_1, \ldots, \beta_s\}. \tag{1}$$

---

*Consider, for example, $s = k + 1 = 3$. Write $p(x) = \prod_{i=1}^{3}(x - \alpha_i) = x^3 - (\alpha_1 + \alpha_2 + \alpha_3)x^2 + (\alpha_1\alpha_2 + \alpha_1\alpha_3 + \alpha_2\alpha_3)x - \alpha_1\alpha_2\alpha_3$, and let $q(x) = \prod_{i=1}^{3}(x - \beta_i) = x^3 - (\beta_1 + \beta_2 + \beta_3)x^2 + (\beta_1\beta_2 + \beta_1\beta_3 + \beta_2\beta_3)x - \beta_1\beta_2\beta_3$. Evidently $a_0 = 1 = b_0$. It is also clear that both $a_1 = -(\alpha_1 + \alpha_2 + \alpha_3)$ equals $b_1 = -(\beta_1 + \beta_2 + \beta_3)$ and $a_2 = \alpha_1\alpha_2 + \alpha_1\alpha_3 + \alpha_2\alpha_3$ equals $b_2 = \beta_1\beta_2 + \beta_1\beta_3 + \beta_2\beta_3$ if and only if $\sum_{i=1}^{3}\alpha_i^j = \sum_{i=1}^{3}\beta_i^j$ for $j = 1, 2$.

For example,

$$\{0,5,5,10\} \stackrel{3}{=} \{1,2,8,9\}.$$

Observe that this 3rd degree solution can be written $\{5,5,10\} \stackrel{3}{=} \{1,2,8,9\}$, as well as $\{0,5,5,10,x\} \stackrel{3}{=} \{1,2,8,9,x\}$ for any integer x. Thus, in the general Tarry-Escott problem, neither distinctness of integers nor equal-sized sets were required.

To explore a number of interesting applications of somewhat restricted solutions of (T), **we shall always assume that a solution of (T) consists of distinct integers** $\{\alpha_1, \ldots, \alpha_s; \beta_1, \ldots, \beta_s\}$. This requires that $s > k$. (If $s \leqslant k$, then $p(x) \equiv q(x)$ and the set of roots $\{\alpha_1, \ldots, \alpha_s\}$ of $p(x)$ is a permutation of the set of roots $\{\beta_1, \ldots, \beta_s\}$ of $q(x)$.) Solutions having $s = k + 1$ are called *ideal*. Since these are probably the most interesting and most useful solutions, they will be the ones stressed here.

There are three simple, but general, theorems that are crucial to solving (T). Our first theorem demonstrates that we can operate on (1) according to the rules of elementary algebra.

Theorem 1. *If* $\{\alpha_1, \alpha_2, \ldots, \alpha_s\} \stackrel{k}{=} \{\beta_1, \beta_2, \ldots, \beta_s\}$, *then* $\{\alpha_1 x + y, \alpha_2 x + y, \ldots, \alpha_s x + y\} \stackrel{k}{=} \{\beta_1 x + y, \beta_2 x + y, \ldots, \beta_s x + y\}$ *for all real numbers* x *and* y.

Proof. For each fixed $j \in \{1, 2, \ldots, k\}$, we see that

$$\sum_{i=1}^{s} (\alpha_i x + y)^j = \sum_{i=1}^{s} \left\{ \sum_{r=0}^{j} \binom{j}{r} \alpha_i^{j-r} x^{j-r} y^r \right\} = \sum_{r=0}^{j} \binom{j}{r} x^{j-r} y^r \left\{ \sum_{i=1}^{s} \alpha_i^{j-r} \right\},$$

which (by hypothesis) equals

$$\sum_{r=0}^{j} \binom{j}{r} x^{j-r} y^r \left\{ \sum_{i=1}^{s} \beta_i^{j-r} \right\} = \sum_{i=1}^{s} \left\{ \sum_{r=0}^{j} \binom{j}{r} \beta_i^{j-r} x^{j-r} y^r \right\} = \sum_{i=1}^{s} (\beta_i x + y)^j.$$

Two solutions of (T) are called *equivalent* if one solution comes from the other as in Theorem 1. This theorem shows that each solution $\{\alpha_1, \alpha_2, \ldots, \alpha_s; \beta_1, \beta_2, \ldots, \beta_s\}$ determines an equivalent "reduced form" solution $\{\alpha_1', \alpha_2', \ldots, \alpha_s'; \beta_1', \beta_2', \ldots, \beta_s'\}$ satisfying $\sum_{i=1}^{s} \alpha_i' = 0 = \sum_{i=1}^{s} \beta_i'$. Moreover, one can easily verify that two solutions are equivalent if and only if their corresponding reduced forms are equivalent.

The following illustration may be useful.

Example 1. (a) To change the solution $\{2,3,7\} \stackrel{2}{=} \{1,5,6\}$ to reduced form, subtract 4 (the average of each side) from each term and obtain $\{-2, -1, 3\} \stackrel{2}{=} \{-3, 1, 2\}$. Here, of course, $x = 1$ and $y = -4$ in Theorem 1.

(b) To change $\{1, 5, 8, 12\} \overset{3}{=} \{2, 3, 10, 11\}$ to reduced form, subtract the mean $13/2$ from each term to get $\{-11/2, -3/2, 3/2, 11/2\} \overset{3}{=} \{-9/2, -7/2, 7/2, 9/2\}$. Since we want integral solutions, we multiply by 2 and obtain the corresponding reduced form $\{-11, -3, 3, 11\} \overset{3}{=} \{-9, -7, 7, 9\}$. This meant taking $x = 2$ and $y = -13$ in Theorem 1.

It will sometimes be useful to express solutions of (T) in compact form. Accordingly, $\pm\{3, 11\} \overset{3}{=} \pm\{7, 9\}$ will denote $\{-11, -3, 3, 11\} \overset{3}{=} \{-9, -7, 7, 9\}$.

Our second theorem shows how to generate solutions of degree $(k + 1)$ from a solution of degree k. Before proceeding, we define the *difference set* $\mathcal{D} = \{|\alpha_i - \alpha_j| \,\|\, \beta_i - \beta_j| : 1 \leqslant i, j \leqslant s\}$ corresponding to the solution $\{\alpha_1, \alpha_2, \ldots, \alpha_s; \beta_1, \beta_2, \ldots, \beta_s\}$ of (T).

Theorem 2. *If $\{\alpha_1, \alpha_2, \ldots, \alpha_s\} \overset{k}{=} \{\beta_1, \beta_2, \ldots, \beta_s\}$ and z is a real number, then*

$$\{\alpha_1, \alpha_2, \ldots, \alpha_s, \beta_1 + z, \beta_2 + z, \ldots, \beta_s + z\}$$
$$\overset{k+1}{=} \{\beta_1, \beta_2, \ldots, \beta_s, \alpha_1 + z, \alpha_2 + z, \ldots, \alpha_s + z\}.$$

Proof. It is not possible that $\beta_i + z = \alpha_j + z$ since this leads to $\beta_i = \alpha_j$. Since $\{\alpha_1, \alpha_2, \ldots, \alpha_s\} \overset{k}{=} \{\beta_1, \beta_2, \ldots, \beta_s\}$ and (by Theorem 1) $\{\alpha_1 + z, \alpha_2 + z, \ldots, \alpha_s + z\} \overset{k}{=} \{\beta_1 + z, \beta_2 + z, \ldots, \beta_s + z\}$, we surely have

$$\sum_{i=1}^{s} \alpha_i^j + \sum_{i=1}^{s} (\beta_i + z)^j = \sum_{i=1}^{s} \beta_i^j + \sum_{i=1}^{s} (\alpha_i + z)^j \quad \text{for} \quad 1 \leqslant j \leqslant k.$$

It remains only to use these k equations and equate like terms in

$$\left(\sum_{i=1}^{s} \alpha_i \right)^{k+1} + \left\{ \sum_{i=1}^{s} (\beta_i + z) \right\}^{k+1} = \left(\sum_{i=1}^{s} \beta_i \right)^{k+1} + \left\{ \sum_{i=1}^{s} (\alpha_i + z) \right\}^{k+1}$$

in order to obtain

$$\sum_{i=1}^{s} \alpha_i^{k+1} + \sum_{i=1}^{s} (\beta_i + z)^{k+1} = \sum_{i=1}^{s} \beta_i^{k+1} + \sum_{i=1}^{s} (\alpha_i + z)^{k+1},$$

and thus complete our proof.

Theorem 2 shows how a particular solution of (T) can be used to build up solutions of any desired degree. If $z \notin \mathcal{D}$, then no equality $\alpha_i = z + \alpha_j$ or $\beta_i + z = \beta_j$ is possible, and an application of Theorem 2 quadruples the size of the solution while merely increasing k by one. Therefore in order to keep down the number of terms, it is advisable to take z as the integer in $\mathcal{D}$ having the greatest frequency. If that integer occurs t times, then t terms in each part of the solution set are identical and may be omitted.

Example 2. To illustrate the power of Theorem 2, we show how $\{1,9\} \overset{1}{=} \{4,6\}$ generates the following solutions of (T):

$$\{1,9\} \overset{1}{=} \{4,6\}$$

$$[z=2] \qquad \{1,8,9\} \overset{2}{=} \{3,4,11\}$$

$$[z=1] \qquad \{1,5,8,12\} \overset{3}{=} \{2,3,10,11\}$$

$$[z=7] \qquad \{1,5,9,17,18\} \overset{4}{=} \{2,3,11,15,19\}$$

$$[z=8] \qquad \{1,5,10,18,23,27\} \overset{5}{=} \{2,3,13,15,25,26\}$$

$$[z=13] \quad \{1,5,10,16,27,28,38,39\} \overset{6}{=} \{2,3,13,14,25,31,36,40\}$$

$$[z=11] \quad \{1,5,10,24,28,42,47,51\} \overset{7}{=} \{2,3,12,21,31,40,49,50\}.$$

The solution $\{1,9\} \overset{1}{=} \{4,6\}$ has difference set $\mathcal{D} = \{8|2\}$. Taking $z = 2$, we obtain $\{1,9,\cancel{6},8\} \overset{2}{=} \{4,\cancel{6},3,11\}$. To apply Theorem 2 again, form the difference set

$$\mathcal{D} = \left\{ \begin{matrix} 7 & 1 \\ & 8 \end{matrix} \ \middle| \ \begin{matrix} 1 & 7 \\ & 8 \end{matrix} \right\}$$

and take $z = 1$ to get $\{1,\cancel{9},8,5,\cancel{4},12\} \overset{3}{=} \{\cancel{4},3,11,2,10,\cancel{9}\}$. The difference set for this solution (reordered by size) is

$$\mathcal{D} = \left\{ \begin{matrix} 4 & 3 & 4 \\ 7 & 7 & \\ 11 & & \end{matrix} \ \middle| \ \begin{matrix} 1 & 7 & 1 \\ 8 & 8 & \\ 9 & & \end{matrix} \right\}.$$

To form a fourth degree solution having as few terms as possible, choose $z = 7$ and obtain $\{1,5,9,17,18\} \overset{4}{=} \{2,3,11,15,19\}$. Another application of Theorem 2, with $z = 8$, produces $\{1,5,10,18,23,27\} \overset{5}{=} \{2,3,13,15,25, 26\}$. (Taking $z = 4$, instead of $z = 8$, yields the nonequivalent solution $\{1,6,7,17,18,23\} \overset{5}{=} \{2,3,11,13,21,22\}$.)

Each of the first five solutions listed above is ideal. However, the integer occurring most frequently in the difference set of our fifth degree solution is 13, and $z = 13$ generates the sixth degree nonideal solution listed. Surprisingly, the difference set for this sixth degree nonideal solution suggests taking $z = 11$, and this leads to the seventh degree ideal solution shown. (This seventh degree solution, in fact, is the seventh degree solution with the smallest integers known.)

Two ideal solutions of (T) are said to *be in sequence* if one of the solutions can be obtained from the other by an application of Theorem 2. It may be observed that for a kth degree solution to be in sequence with a $(k + 1)$st degree solution, the value of z must occur exactly k times in the difference set of the kth degree solution. Thus, the higher the degree k, the less likely it appears that one can find ideal solutions in sequence. Actually, it was shown in 1947 [5] (and verified more recently by using computers [2]) that no sequence of ideal solutions beginning with the first degree extends beyond the fifth degree, and that the only three nonequivalent fifth degree solutions which exist can be reached in this manner. Two of them have already been given in Example 2; the third

$$\{1, 7, 9, 24, 26, 32\} \overset{5}{=} \{2, 4, 12, 21, 29, 31\}$$

may be obtained from $\{1, 5\} \overset{1}{=} \{2, 4\}$ by successively taking $z = 4, 7, 5, 11$.

The sixth degree ideal solution having smallest integers known can be found by starting with $\{1, 12, 20, 23, 31, 42\} \overset{1}{=} \{15, 16, 17, 26, 27, 28\}$ and successively using $z = 11, 1, 19, 17, 13$ to obtain

$$\{1, 19, 28, 59, 65, 90, 102\} \overset{6}{=} \{2, 14, 39, 45, 76, 85, 103\}.$$

The last three solutions in this process are seen to be ideal solutions in sequence.

To find the eighth degree ideal solution having the smallest integers known, start with $\{1, 10, 12, 34, 35, 57, 59, 68\} \overset{1}{=} \{20, 22, 29, 31, 38, 40, 47, 49\}$ and successively use $z = 9, 1, 11, 17, 23, 29, 41$ to obtain

$$\{1, 25, 31, 84, 87, 134, 158, 182, 198\} \overset{8}{=} \{2, 18, 42, 66, 113, 116, 169, 175, 199\}.$$

The last two solutions in this process form a sequence of ideal solutions.

A ninth degree ideal solution was reported in 1944 (see [10], 54–56) in which all but two of twenty members are four or five digit numbers.

To date, it is not known whether or not there exist ideal solutions of degree $k > 9$. Recent attempts to find solutions of arbitrary degree $k > 0$ began by defining the function $s(k)$ as the least number s such that $\{\alpha_1, \alpha_2, \ldots, \alpha_s\} \overset{k}{=} \{\beta_1, \beta_2, \ldots, \beta_s\}$ has solutions. We already know (from Example 2 and the preceding remarks) that $s(k) = k + 1$ for $1 \leq k \leq 9$. Using an IBM 1620 computer, Barrodale [2] obtained the following:

k	10	11	12	13	14	15	16	17	18	19	20	21	22
$s(k)$	14	18	24	30	30	30	38	48	58	58	65	80	84

In correspondence with E. B. Escott (of the Tarry-Escott duo) it was learned that Theorem 2 can be reversed to produce "predecessors," instead of "successors," of a given solution.

Theorem 3.   *Suppose* $\{\alpha_1, \alpha_2, \ldots, \alpha_s\} \overset{k}{=} \{\beta_1, \beta_2, \ldots, \beta_s\}$, *where* $\alpha_i < \beta_i$ $(1 \leq i \leq r)$ *and* $\beta_i < \alpha_i$ $(r + 1 \leq i \leq s)$ *satisfy* $\alpha_i \equiv \beta_i \pmod{d}$ *for* $1 \leq i \leq s$. *Then* $\{\alpha_1, \alpha_1 + d, \alpha_1 + 2d, \ldots, \beta_1 - d; \alpha_2, \alpha_2 + d, \ldots, \beta_2 - d; \ldots,;$ $\alpha_r, \alpha_r + d, \ldots, \beta_r - d\} \overset{k-1}{=} \{\beta_{r+1}, \beta_{r+1} + d, \ldots, \alpha_{r+1} - d; \beta_{r+2}, \beta_{r+2} + d, \ldots, \alpha_{r+2} - d; \ldots; \beta_s, \beta_s + d, \ldots, \alpha_s - d\}.$

The mechanics of Theorem 3 can be illustrated as follows:

Example 3. Construct a predecessor for the ideal solution

$$\{1, 8, 15, 29, 36, 43\} \overset{5}{=} \{3, 4, 21, 23, 40, 41\}. \tag{2}$$

To find a suitable value for d, make up a table of absolute differences as follows:

α \ β	3	4	21	23	40	41
1	2	3	20	22	39	40
8	5	4	13	15	32	33
15	12	11	6	8	25	26
29	26	25	8	6	11	12
36	33	32	15	13	4	5
43	40	39	22	20	3	2

The integers $2, 3, 5, 11, 13$ appear in each row of the table and thus are candidates for d. Suppose we consider $d = 13$. Having chosen $d = 13$, it is now possible to label the α's and β's so that $\alpha_i \equiv \beta_i \pmod{13}$ for $i = 1, 2, \ldots, 6$.

α \ β	$\beta_4 \\ \| \\ 3$	$\beta_6 \\ \| \\ 4$	$\beta_2 \\ \| \\ 21$	$\beta_5 \\ \| \\ 23$	$\beta_1 \\ \| \\ 40$	$\beta_3 \\ \| \\ 41$
$\alpha_1 = 1$					39	
$\alpha_2 = 8$			13			
$\alpha_3 = 15$						26
$\alpha_4 = 29$	26					
$\alpha_5 = 36$				13		
$\alpha_6 = 43$		39				

Here $r = 3$, and a predecessor to (2) is constructed by the schematic

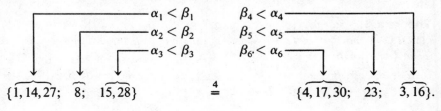

$$\{\widehat{1, 14, 27};\ \ \widehat{8};\ \ \widehat{15, 28}\} \overset{4}{=} \{\widehat{4, 17, 30};\ \ \widehat{23};\ \ \widehat{3, 16}\}.$$

Thus

$$\{1, 8, 14, 15, 27, 28\} \overset{4}{=} \{3, 4, 16, 17, 23, 30\}. \tag{3}$$

It can be shown, by checking the other values $d = 2, 3, 5, 11$, that no predecessor to (2) is an ideal solution.

The reader may also wish to apply Theorem 3 to (3), and successively use $d = 2, 11, 8$ to generate a sequence of ideal solutions descending down to $\{1, 9\} \overset{1}{=} \{4, 6\}$.

Applications. Many applications of the Tarry-Escott problem are known ([1], [7]). A few of these, that the reader may wish to explore in greater detail, are described briefly below.

1.   Multiple unit roots of certain equations. In order that $x = 1$ be a root of multiplicity 4 of

$$f(x) = x^{12} - x^{11} - x^{10} + x^8 + x^5 - x^3 - x^2 + x = 0,$$

we need

$$f(1) = f'(1) = f''(1) = f'''(1) = 0.$$

Clearly, $f(1) = 0$. And $f'(1) = 0$ holds since $12 + 8 + 5 + 1 = 11 + 10 + 3 + 2$. Similarly, $f''(1) = 0$ since $12^2 + 8^2 + 5^2 + 1^2 = 11^2 + 10^2 + 3^2 + 2^2$, and $f'''(0) = 0$ follows from $12^3 + 8^3 + 5^3 + 1^3 = 11^3 + 10^3 + 3^3 + 2^3$. Thus $f(x) = (x - 1)^4 g(x)$, for some eighth degree polynomial $g(x)$, since the exponents characterizing $f(x)$ satisfy: $\{1, 5, 8, 12\} \overset{3}{=} \{2, 3, 10, 11\}$.

Escott once indicated (in private correspondence) that solutions of (T) could be used to form factorable polynomials of the above type. For instance, the listed solutions in Example 2 produce:

$$x^9 - x^6 - x^4 + x = (x^6 - x)(x^3 - 1)$$
$$x^{11} - x^9 - x^8 + x^4 + x^3 - x = (x^6 - x)(x^3 - 1)(x^2 - 1)$$
$$x^{12} - x^{11} - x^{10} + x^8 + x^5 - x^3 - x^2 + x = (x^6 - x)(x^3 - 1)(x^2 - 1)(x - 1)$$
$$x^{19} - x^{18} - x^{17} + x^{15} + x^{11} - x^9 - x^5 + x^3 + x^2 - x = (x^6 - x)(x^3 - 1)(x^2 - 1)(x - 1)(x^7 - 1)$$

and so on. Note that all the additional factors, after the first equation, are of the form $x^z - 1$.

2.   Reducibility of certain polynomials. In 1933–1935, it was shown [8], [6] that for a prime number p, polynomials of the form

$$p(x) = a(x - x_1)(x - x_2) \cdots (x - x_n) \pm p \quad (x_i \neq x_j) \tag{4}$$

are irreducible in the rational domain for odd $n > 3$, and may consist only of two factors of degree $n/2$ when $n \geqslant 8$ is even. The paper [6] established that:

Polynomials (4) having degree $n = 2m \geqslant 10$ are reducible if and only if

$$\sqrt{a}\,(x_i - x_1)(x_i - x_2) \cdots (x_i - x_m) = |p \pm 1| \quad \text{for each} \atop i = m + 1, \ldots, n \tag{5}$$

and the decomposition can be made in any one of the following four equivalent forms:

$$\left[\sqrt{a}\,(x-x_1)\cdots(x-x_m)\pm 1\right]\left[\sqrt{a}\,(x-x_1)\cdots(x-x_m)\pm p\right]$$

$$\left[\sqrt{a}\,(x-x_1)\cdots(x-x_m)\pm 1\right]\left[\sqrt{a}\,(x-x_{m+1})\cdots(x-x_n)\mp 1\right]$$

$$\left[\sqrt{a}\,(x-x_{m+1})\cdots(x-x_n)\mp p\right]\left[\sqrt{a}\,(x-x_1)\cdots(x-x_m)\pm p\right]$$

$$\left[\sqrt{a}\,(x-x_{m+1})\cdots(x-x_n)\mp p\right]\left[\sqrt{a}\,(x-x_{m+1})\cdots(x-x_n)\mp 1\right]$$

with proper attention to sign.

The reducibility conditions (5) were also shown to be equivalent to finding an $(m-1)$th degree ideal solution $\{x_1,\ldots,x_m\}\overset{m-1}{=}\{x_{m+1},\ldots,x_n\}$ for (T). (An additional condition determines the primes p for which the decomposition exists.) By way of illustration, the ideal solution $\{1,6,7,17,18,23\}\overset{5}{=}\{2,3,11,13,21,22\}$, whose reduced form is $\pm\{5,6,11\}\overset{5}{=}\pm\{1,9,10\}$, gives rise to the following twelfth degree polynomial factorization:

$$(x^2-1)(x^2-5^2)(x^2-6^2)(x^2-9^2)\,(x^2-10^2)(x^2-11^2)+100799$$
$$=\left[(x^2-1)(x^2-9^2)(x^2-10^2)-100799\right]$$
$$\times\left[(x^2-5^2)(x^2-6^2)(x^2-11^2)+100799\right].$$

The ideal solution $\{1,5,8,12\}\overset{3}{=}\{2,3,10,11\}$ gives rise to the eighth degree polynomial factorization:

$$(x-1)(x-2)(x-3)(x-5)(x-8)(x-10)(x-11)(x-12)+179$$
$$=\left[(x-1)(x-5)(x-8)(x-12)+179\right]$$
$$\times\left[(x-2)(x-3)(x-10)(x-11)-179\right].$$

General ideal solutions for this reducibility problem have been found [3, pp. 49–58] for $m=3,4$. Special solutions and numerical examples have also been found ([4], 705–713, [10]) for $m=5,6,7,8,9,10$. Thus, reducible polynomials (4) can be found for $n=10,12,14,16,18,20$. The existence or nonexistence of such reducible polynomials having degree $n=2m\geqslant 22$ is unanswered.

3. Ideal solutions yield magic-type squares.

The existence of an ideal solution $\{a_1,\ldots,a_{k+1}\}\overset{k}{=}\{b_1,\ldots,b_{k+1}\}$ for $k\geqslant 3$ is equivalent (see the discussion in Section 2) to

$$\sqrt{a}\,(b_i-a_1)(b_i-a_2)\cdots(b_i-a_{k+1})=|p\pm 1|\quad\text{for each}$$
$$i=1,2,\ldots,k+1.$$

Clearly a must be a perfect square, and $\sqrt{a}$ must divide $|p\pm 1|$. Assume that $|p \pm 1| = \sqrt{a} \cdot d$ so that

$$(b_i - a_1)(b_i - a_2) \cdots (b_i - a_{k+1}) = d \quad \text{for each} \quad i = 1, 2, \ldots, k+1.$$

Therefore, in the square

$$
\begin{array}{cccc}
|b_1 - a_1| & |b_1 - a_2| & \cdots & |b_1 - a_{k+1}| \\
|b_2 - a_1| & |b_2 - a_2| & \cdots & |b_2 - a_{k+1}| \\
\vdots & \vdots & \cdots & \vdots \\
|b_{k+1} - a_1| & |b_{k+1} - a_2| & \cdots & |b_{k+1} - a_{k+1}|,
\end{array}
$$

each product of the elements of the rows is equal to d. And it can also be shown that the products of the column elements also have value d. The above square is almost a "multiplication magic square"; it is not one because the products of the elements of the principal diagonals do not have this property.

The above square is "centrally skew-symmetric" if k is odd, and "axi-symmetric" if k is even. For example, $\{1, 5, 8, 12\} \overset{3}{=} \{2, 3, 10, 11\}$ yields

$$
\begin{array}{cccc}
1 & 3 & 6 & 10 \\
2 & 2 & 5 & 9 \\
9 & 5 & 2 & 2 \\
10 & 6 & 3 & 1,
\end{array}
$$

whereas $\{1, 5, 9, 17, 18\} \overset{4}{=} \{2, 3, 11, 15, 19\}$ produces

$$
\begin{array}{ccccc}
1 & 3 & 7 & 15 & 16 \\
2 & 2 & 6 & 14 & 15 \\
10 & 6 & 2 & 6 & 7 \\
14 & 10 & 6 & 2 & 3 \\
18 & 14 & 10 & 2 & 1.
\end{array}
$$

Another, more recent article [1], shows how magic cubes are related to the Tarry-Escott problem.

REFERENCES

1. A. Adler and R. Li Shuo-Yen, Magic cubes and Prouhet sequences, Amer. Math. Monthly, 84 (1977) 618–627.
2. I. Barrodale, A note on equal sums of like powers, Math. Comp., 20 (1966) 318–322.
3. L. E. Dickson, Introduction to the Theory of Numbers, Univ. of Chicago Press, Chicago, 1929.
4. ———, History of the Theory of Numbers, vol. 2, Chelsea, New York, 1920.
5. H. L. Dorwart, Sequences of ideal solutions in the Tarry-Escott problem, Bull. Amer. Math. Soc., 53 (1947) 381–391.
6. ———, Concerning certain reducible polynomials, Duke Math. J., 1 (1935) 70–73.
7. ——— and O. E. Brown, The Tarry-Escott problem, Amer. Math. Monthly, 44 (1937) 613–626.
8. ——— and O. Oystein, Criteria for the reducibility of polynomials, Ann. of Math., 34:2 (1933) 81–94. Also, 35:2 (1934) 195.
9. E. B. Escott, Logarithmic series, Quart. J. Pure and Applied Math., 41 (1910) 141–167.
10. A. Gloden, Mehrgradige Gleichungen, Noordhoff, Groningen, Holland, 1944.
11. G. Tarry, L'intermédiaire des mathématiciens, 19 (1912) 219–221.

The Digital Root Function

Thomas P. Dence, *Bowling Green State University, Huron, OH*

While teaching a course called Topics in Modern Mathematics, I became acquainted with the digital root function and its use in determining whether or not a large number was a perfect square or cube. Digital roots are also used in the analysis of card tricks and games [1] as well as in the study of self-numbers (see [3], for example). There are many ways in which students can "play" with digital roots. A study of the digital root function, using a computer to gather data, provides excellent motivation and exercises for students to learn mathematics and programming.

I. Basic Properties

The digital root function $y = DR(x)$ simply maps the positive integers onto the set of digits $\{1, 2, \ldots, 9\}$ by assigning to each integer that digit obtained by continually adding the digits until a single digit is reached. Thus $DR(13) = 4$ and $DR(93) = 3$. An elementary program designed to determine the digital root of the first 100 integers quickly shows (see Table 1) that the function appears to be periodic with a period of 9. Thus two integers differing by a multiple of 9 have the same digital root. Accordingly, the digital root of a multiple of 9 is always 9.

It is pointed out in [1] that the digital root of an integer is the remainder that is left after division by 9. This process (known as "casting out nines") always works if we restrict the remainder to the set $\{1, 2, \ldots, 9\}$. To corroborate what Table 1 suggests, we prove

Theorem 1. *If $n = 9q + r$ ($1 \leqslant r \leqslant 9$), then $DR(n) = r$. Therefore,*

$$DR(n + 9k) = DR(n) \quad and \quad DR(9k) = 9 \tag{1}$$

for all natural numbers n, k.

Proof. The assertions in (1) are immediate consequences of the preceding property of DR. Indeed,

$$DR(n + 9k) = DR(9q + r + 9k) = DR\big[9(q + k) + r\big] = r = DR(n).$$

Then $DR(9k) = DR[9 + 9(k - 1)] = DR(9) = 9.$

Table 1.

n	$DR(n)$	n^2	$DR(n^2)$	n^3	$DR(n^3)$	n^4	$DR(n^4)$
1	1	1	1	1	1	1	1
2	2	4	4	8	8	16	7
3	3	9	9	27	9	81	9
4	4	16	7	64	1	256	4
5	5	25	7	125	8	625	4
6	6	36	9	216	9	1296	9
7	7	49	4	343	1	2401	7
8	8	64	1	512	8	4096	1
9	9	81	9	729	9	6561	9
10	1	100	1	1000	1	10000	1
11	2	121	4	1331	8	14641	7
12	3	144	9	1728	9	20736	9
13	4	169	7	2197	1	28561	4
14	5	196	7	2744	8	38416	4
15	6	225	9	3375	9	50625	9
⋮	⋮	⋮	⋮	⋮	⋮	⋮	⋮
100	1	10000	1	1000000	1	100000000	1

The main result, that $DR(n) = r$ for $n = 9q + r$, is proved by induction on the number of digits that n has. Evidently $DR(n) = n$ if n is a one-digit natural number. Suppose $n = d_2 d_1$ is a two-digit number. Then $n = 10d_2 + d_1 = 9d_2 + (d_1 + d_2)$ with $1 \leqslant d_1 + d_2 \leqslant 18$. If $1 \leqslant d_1 + d_2 \leqslant 9$, then $DR(n) = DR(d_1 + d_2) = d_1 + d_2$. If $10 \leqslant d_1 + d_2 \leqslant 18$, then $d_1 + d_2 = 9(1) + r$. Therefore, $DR(n) = DR(d_1 + d_2) = DR(9 + r) = DR[10 + (r - 1)] = r$, and this is the remainder in the representation $n = 9d_2 + (d_1 + d_2) = 9(d_2 + 1) + r$. Finally, assume that our result holds for k-digit numbers and let $n = d_{k+1} d_k \ldots d_2 d_1$. Then

$$n = \sum_{i=2}^{k+1} (10^{k-1} - 1)x_k + \sum_{i=1}^{k+1} x_i = 9 \sum_{i=2}^{k+1} \left(\frac{10^{k-1} - 1}{9} \right) x_i + \sum_{i=1}^{k+1} x_i,$$

where $1 \leqslant \sum_{i=1}^{k+1} x_i \leqslant 9(k+1)$. If $1 \leqslant \sum_{i=1}^{k+1} x_i \leqslant 9$, then $DR(n) = \sum_{i=1}^{k+1} x_i$ agrees with the above remainder. If $10 \leqslant \sum_{i=1}^{k+1} x_i \leqslant 9(k+1)$, then $\sum_{i=1}^{k+1} x_i = 9(x) + r$ for $1 \leqslant x \leqslant k$. Since $9(k+1) < 10^{k-1}$ (that is, $\sum_{i=1}^{k+1} x_i$ has less than $k + 1$ digits), our induction hypothesis yields $DR(n) = DR(\sum_{i=1}^{k+1} x_i) = r$, which is the remainder in $n = 9 \left[\sum_{i=2}^{k+1} ((10^{k-1} - 1)/9)x_i + x \right] + r$.

Remark. This process generalizes to bases other than 10. If a number x in base 10 is written, say $x_k = abc$ in base k, then $x_k \equiv (a + b + c) \pmod{k - 1}$. See [1] for a proof.

We have already observed that $DR(n) \in \{1, 2, \ldots, 9\}$. Therefore, $DR\{DR(n)\} = DR(n)$. A few additional properties of DR are summarized below.

Theorem 2.   The digital root function satisfies:
   (i) $DR(n) = DR(m) \Leftrightarrow |m - n| = 9k \ (k \geqslant 0)$.
   (ii) $DR(n + N) = DR(m + N) \Leftrightarrow DR(n) = DR(m)$.
   (iii) $DR(n) = DR(m) \Rightarrow DR(nN) = DR(mN)$. *The converse also holds for* $N \nmid 9$.
   (iv) $DR(n + m) = DR[DR(n) + DR(m)]$.
   (v) $DR(n \cdot m) = DR[n \cdot DR(m)] = DR[m \cdot DR(n)]$.

Proof.   (i) If $m = n + 9k$, then $DR(n) = DR(m)$ by (1). Suppose conversely that $n < m$, and write $m = 9p + s$ and $n = 9q + r$. Then $s = DR(m) = DR(n) = r$, and so $m - n = 9(p - q) > 0$. For (ii), first consider $DR(n) = DR(m)$. Evidently $DR(n + N) = DR(m + N)$ if $n = m$. If $n < m$, then [by (i)] $m = n + 9k$ and $DR(m + N) = DR[(n + N) + 9k] = DR(n + N)$. In the other direction, suppose $DR(n + N) = DR(m + N)$ and let $n < m$. Then $m + N = n + N + 9k$, and so $m = n + 9k$ yields $DR(m) = DR(n)$. Both assertions in (iii) are obviously true for $n = m$. If $DR(n) = DR(m)$ and $n < m$, then [by (i)] $m = n + 9k$, and $mN = nN + 9(kN)$ yields $DR(mN) = DR(nN)$. Conversely, $DR(nN) = DR(mN)$ for $n < m$ implies that $mN = nN = 9k$. Since $n | 9k$ and $n \nmid 9$, it is clear that $m = N + 9(k/N)$ and $DR(m) = DR(n)$. The proof of (iv) is simple: If $DR(n) = n'$ and $DR(m) = m'$, then $n = 9j + n'$ and $m = 9k + m'$. Therefore, $DR(n + m) = DR[9j + n' + 9k + m'] = DR[(n' + m') + 9(j + k)] = DR(n' + m') = DR[DR(n) + DR(m)]$. It is easily verified that (iv) extends to finite sums

$$DR(n_1 + n_2 + \cdots + n_m)$$
$$= DR[DR(n_1) + DR(n_2) + \cdots + DR(n_m)].$$

This, of course, leads to (v).

The digital root function also has the following surprising property.

Corollary 3. $DR(n \cdot m) = DR[DR(n) \cdot DR(m)]$ *and* $DR(n^k) = DR[\{DR(n)\}^k]$.

Proof. $DR(n \cdot m) = DR[n \cdot DR(m)] = DR[n + n + \cdots + n]$ (with $DR(m)$ summands), and this equals $DR[DR(n) + DR(n) + \cdots + DR(n)] = DR[DR(n) \cdot DR(m)]$. The exponential property follows by induction.

Interested students can explore other related properties of the digital root function such as $DR[(n + 9)^k] = DR(n^k)$ and $DR[(n + 3)^3] = DR(n^3)$. Why is $DR[(n + 3)^k \neq DR(n^k)$ in general? And, what can be said about division properties of digital roots?

II. Powers and Primes

One application of the digital root function is in determining whether or not a number is a perfect power. Table 1 showed that a number is a perfect square only if its digital root is 1, 4, 7, or 9. It is immediately clear, from the $DR(n^2)$ columns of Table 1, that 123456789123 is not a square since its digital root is 6. It is already known, by (i), that $DR(n^m)$ has period 9 for all integral powers $m > 0$. Using this and Theorem 2 (iv), we can write a program which computes (quite rapidly) the digital roots of higher powers.

Table 2.

n	DR (n)	DR (n^2)	DR (n^3)	DR (n^4)	DR (n^5)	DR (n^6)	DR (n^7)	DR (n^8)	DR (n^9)	DR (n^{10})	DR (n^{11})
1	1	1	1	1	1	1	1	1	1	1	1
2	2	4	8	7	5	1	2	4	8	7	5
3	3	9	9	9	9	9	9	9	9	9	9
4	4	7	1	4	7	1	4	7	1	4	7
5	5	7	8	4	2	1	5	7	8	4	2
6	6	9	9	9	9	9	9	9	9	9	9
7	7	4	1	7	4	1	7	4	1	7	4
8	8	1	8	1	8	1	8	1	8	1	8
9	9	9	9	9	9	9	9	9	9	9	9

Table 2 provides some interesting observations which could (and should!) be easily verified by students. A partial sample includes:

(a) If $3|m$, then $DR(n^m)$ has period 3.
(b) $DR(3^k) = DR(6^k) = DR(9^k) = 9$ for $k > 1$.
(c) $DR(n^k) = DR(n^{k+6})$ for $k > 1$.
(d) $DR(n^k) \neq 3$ or 6 for $k > 1$.

(e)
$$DR(n^{6k}) = \begin{cases} 1, & \text{if } n \neq 3m, \\ 9, & \text{if } n = 3m. \end{cases}$$

(f)
$$DR(n^{6k-3}) = \begin{cases} 1, & \text{if } n = 3m + 1, \\ 8, & \text{if } n = 3m + 2, \\ 9, & \text{if } n = 3m. \end{cases}$$

(g) Let $k(n)$ denote the proportion of times that the integer k occurs in the first n columns. Then

$$\lim_{n \to \infty} k(n) = \begin{cases} 15/54, & k = 1, \\ 2/54, & k = 2 \text{ or } 5, \\ 0, & k = 3 \text{ or } 6, \\ 6/54, & k = 4 \text{ or } 7, \\ 5/54, & k = 8, \\ 18/54, & k = 9. \end{cases}$$

We now turn our attention to the primes in the hope that some revealing property could be unearthed. This served as good motivation for students to become involved with the programming exercises of economically computing the primes and their digital roots. Table 3 lists the first 500 primes, together with a periodic tally of the occurrences of the digital roots. A listing of twin primes (i.e., consecutive primes p, $p + 2$) with their digital roots and occurrences is supplemental.

Table 3.

n	prime	DR(prime)	twin primes	dig. roots	occurrence
1	2	2			
2	3	3			
3	5	5	3, 5	3, 5	1
4	7	7	5, 7	5, 7	1
5	11	2			
6	13	4	11, 13	2, 4	1
7	17	8			
8	19	1	17, 19	8, 1	2
9	23	5			
10	29	2			
11	31	4	29, 31	2, 4	2
12	37	1			
13	41	5			
⋮					
50	229	4	227, 229	2, 4	6

the number of times 1 has occurred so far as a digital root is 8
the number of times 2 has occurred so far as a digital root is 10
the number of times 3 has occurred so far as a digital root is 1
the number of times 4 has occurred so far as a digital root is 9
the number of times 5 has occurred so far as a digital root is 8
the number of times 6 has occurred so far as a digital root is 0
the number of times 7 has occurred so far as a digital root is 7
the number of times 8 has occurred so far as a digital root is 7

51	233	8
52	239	5
⋮		
100	541	1

the number of times 1 has occurred so far as a digital root is 16
⋮
the number of times 8 has occurred so far as a digital root is 16

500	3571	7
⋮		

the number of times 8 has occurred so far as a digital root is 81.

The occurrence of the digital roots of the primes is somewhat noteworthy because of the near-equal distribution of the digits $1, 2, 4, 5, 7, 8$. Other than one particular exception [DR(3) = 3], no other digits are possible digital roots of primes since $3 \nmid \text{DR}(n)$ for a prime number $n \neq 3$. This distribution is summarized as follows.

First n primes	Frequency of Occurrence					
	DR( ) = 1	DR( ) = 2	DR( ) = 4	DR( ) = 5	DR( ) = 7	DR( ) = 8
50	8	10	9	8	7	7
100	16	18	15	18	16	16
150	23	26	25	25	24	26
200	32	36	32	33	32	34
250	39	44	42	44	39	41
300	46	51	50	53	50	49
350	57	61	56	60	58	57
400	67	72	63	68	65	64
450	74	80	73	76	72	74
500	82	88	83	84	81	81

If we conjecture that these 6 digits occur with equal probability of 1/6 and if we run a chi-squared test on it with $n = 500$, we find that chi-square $= .42$. Therefore our data is "too good" at the .01 level of significance, and we accept the hypothesis. Also surprising is the fact that the distributions remain quite steady throughout. It is strange that the digit 2 always slightly dominates in occurrence.

If the primes are listed according to their digital roots as shown below

DR( ) = 1	DR( ) = 2	DR( ) = 4	DR( ) = 5	DR( ) = 7	DR( ) = 8
19	2	13	5	7	17
37	11	31	23	43	53
73	29	67	41	61	71
109	47	103	59	79	89
127	83	139	113	97	107
⋮	⋮	⋮	⋮	⋮	⋮

we again find a near-equal distribution in the occurrence of the differences of the primes in each column. All but one of the differences are multiples of 18. For the first 500 primes, we have

Difference	Frequency of Occurrence					
	column 1	column 2	column 3	column 4	column 5	column 6
9	0	1	0	0	0	0
18	31	32	30	30	34	32
36	23	24	23	21	16	20
54	13	15	16	17	14	11
72	7	10	7	7	5	12
90	3	3	2	5	4	3
108	2	0	1	1	2	0
126	2	0	1	2	3	0
144	0	1	1	0	1	1
162	0	1	1	0	1	1

Table 4.

Frequency of occurrence (b = base, k = digital root).

k \ b	1	2	3	4	5	6	7	8	9	10	11	12	13	14	15	16	17	18	19	20	21	22	23	24	25	26	27	28	29
4	195	204	1																										
5	195	1	204	0																									
6	98	102	103	96	1																								
7	195	1	1	0	203	0																							
8	65	66	67	67	67	67	1																						
9	93	1	102	0	102	0	102	0																					
10	67	66	1	63	68	0	67	68	0																				
11	98	1	103	0	1	0	99	0	98	0																			
12	39	40	40	38	39	40	40	39	41	43	1																		
13	93	1	1	0	102	0	102	0	0	0	101	0																	
14	32	34	34	30	34	34	33	33	35	34	33	33	1																
15	65	1	67	0	67	0	1	0	65	0	67	0	67	0															
16	47	51	1	49	1	0	49	51	0	0	51	0	51	53	0														
17	45	1	51	0	51	0	48	0	48	0	51	0	51	0	54	0													
18	25	26	28	23	22	22	26	27	24	25	27	28	23	26	23	24	1												
19	67	1	1	0	68	0	65	0	0	0	71	0	63	0	0	0	64	0											
20	20	21	25	26	23	21	21	21	22	22	23	21	25	23	20	21	21	23	1										
21	46	1	50	0	1	0	52	0	46	0	52	0	53	0	0	0	49	0	50	0									
22	31	34	1	32	33	0	1	34	0	35	32	0	38	0	0	33	33	0	40	40	0								
23	39	1	40	0	39	0	42	0	38	0	1	0	42	0	21	0	40	0	40	0	40	0							
24	19	18	19	19	17	16	17	15	17	19	18	16	16	18	21	20	19	19	21	18	20	17	1						
25	43	1	1	0	52	0	51	0	0	0	50	0	50	0	0	0	50	0	51	0	0	0	51	0					
26	19	20	19	20	1	22	20	22	22	0	20	21	21	20	0	17	17	19	20	0	16	18	20	22	0				
27	32	1	34	0	34	0	37	0	35	0	33	0	1	0	34	0	30	0	32	0	33	0	31	0	31	0			
28	23	26	1	20	23	0	21	21	0	21	22	0	23	23	0	22	24	0	23	24	0	20	22	0	22	19	0		
29	32	1	36	0	33	0	1	0	32	0	36	0	36	0	33	0	31	0	34	0	0	0	33	0	31	0	31	0	
30	13	16	17	15	15	12	15	15	10	16	15	16	14	14	17	14	12	15	15	16	14	14	14	11	14	12	15	13	1

A difference of 18 occurs about 38% of the time for each different digital root; 36 occurs about 26% of the time; 54 about 17%, and 72 about 9%.

Students can readily verify that twin primes occur only when the corresponding digital roots are one of the pairs $(2, 4)$, $(5, 7)$, or $(8, 1)$. The converse is not true since the consecutive (non-twin) primes 3413 and 3433 have digital roots 2 and 4 respectively.

Our last result illustrates another surprising property of digital roots—namely, the distribution of digital roots of primes when different bases b $(1 \leqslant b \leqslant 30)$ are considered. Table 4 exhibits the occurrence of the digital root k $(k \leqslant b - 1)$ for the first 400 primes.

Table 4 offers more food for thought. Astute students will readily take note of, and may wish to delve further into, the following observations:

(a) In base $2k + 1$, the digit $2k$ occurs zero times. In base $2k$, the digit $2k - 1$ occurs zero or one times.

(b) A prime digit p occurs just once in base $np + 1$ for all $n \geqslant 1$.

(c) In an odd base, the digital roots are concentrated on the odd digits.

REFERENCES

1. Martin Gardner, The Second Scientific American Book of Mathematical Puzzles and Diversions, Simon and Schuster, New York, 1961, pp. 43–50.
2. ———, Mathematics Magic and Mystery, Dover, New York, 1956.
3. ———, Sci. Amer., April 1975, 133.

4. CALCULUS

Historically, calculus developed as two separate branches of mathematics, each of which can be traced back to a fundamental underlying problem.

The impetus for the development of integral calculus began with the problem of calculating the area contained within a closed curve. Thus, Archimedes' methods of exhaustion and infinitesimal mensurations (i.e., geometry applied to the computation of lengths, areas, volumes) may be viewed as the infancy of integral calculus.

About two thousand years later, problems concerning instantaneous rates of change came into prominence. Descartes' invention of coordinate geometry, in 1637, stimulated thinking on how to draw tangents to points on a curve. And in 1638, Galileo's treatise, on the motion of freely falling bodies, drew attention to calculating velocities and accelerations at an instant. These problems sparked the inception of differential calculus.

The first systematic treatment of calculus, published by L'Hospital in 1696, contained the postulate that two quantities which differ by an infinitely small quantity may be taken as equal. Apparently neither Newton nor Leibnitz had successfully resolved the difficulties associated with the concept of "limit." Thus, it is clear that the early formulations of calculus were neither always exact nor always mathematically rigorous. Only recently, after centuries of development, has calculus come to rest on a solid, rigorous mathematical foundation. Today, calculus permeates all branches of science, and it is widely taught both for its intrinsic beauty and for its applicability elsewhere.

The articles in this chapter follow in the spirit of calculus's historical development. They can be used to stimulate interest and provide further insights into the cumulative, evolutionary nature of the subject.

Kenneth Goldberg's article "Does the Formula for Arc Length Measure Arc Length?" correctly points out that most elementary calculus texts contain a logical gap in their development of the formula for arc length. This gap has surprisingly eluded a large part of the mathematical community despite our reliance on mathematical precision. To remove this gap, Warren Page returns to Archimedes. Page's "The Formula for Arc Length *Does* Measure Arc Length" more than resolves Goldberg's critical questions; it demonstrates how deeply ingrained earliest notions (and difficulties?) are in our present developments, and it illustrates how new inexactitudes may evolve by overlooking earlier fundamentals.

Charles Weaver's "Rational Solutions of $x^y = y^x$" and Sidney Kung's "A Vectorial Approach to 'The Chase'" nicely illustrate how far the earliest problems, of finding tangents and calculating instantaneous velocities, have come. Weaver uses calculus to graph $t/\ln t$, and thus obtain a result in number theory. Kung uses vector calculus (a sort of generalized calculus of the real numbers) to solve a cat-and-mouse pursuit problem.

Learning in calculus proceeds in stages (not unlike the subject's evolutionary development). Many elementary calculus concepts are first taught by relying heavily on intuition and other plausibility arguments. As students' understanding of these concepts matures, we can build on and strengthen their previous knowledge by emphasizing the finer aspects of the subject that may not have originally been considered or mastered. Michael Ecker's "Discontinuity Amidst Differentiability" is a good case in point. His article, illustrating unusual mixtures of continuity, discontinuity, and differentiability that functions may enjoy, can be effectively used to sharpen understanding and deepen appreciation of such basic calculus notions.

Does the Formula for Arc Length Measure Arc Length?

Kenneth P. Goldberg, *New York University, NY*

The standard development of the arc length formula employed in most elementary calculus texts is incomplete. There is, in fact, a gap in the usual development of this formula that is glossed over in most texts. This will become clear once we review the methods for computing planar areas and arc lengths.

Evaluating Areas in the Plane

The method of exhaustion was employed by the classical Greek geometers for the purpose of approximating planar areas. As the name implies, the area to be exhausted was literally "exhausted" by a sequence of inscribed figures (such as regular polygons) until a desired degree of accuracy was reached. Figure 1, for example, illustrates how the area inside the circle is approximated by inscribed regular polygons of $n = 3$, $n = 6$, and $n = 12$ sides, respectively. The shaded area (i.e., the error of approximation) can be shown to decrease monotonically to zero as n increases to infinity. In this sense, the area to be approximated is exhausted.

 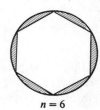 

$n = 3$ $n = 6$ $n = 12$

Figure 1.

The techniques of "sandwiching" the desired measurement between two sequences which converge to the same limit is often employed with, or in place of, the method of exhaustion. This is the familiar technique for evaluating the area between the x-axis and the curve $y = f(x)$ from $x = a$ to $x = b$. Here (Figure 2) one divides $[a, b]$ into n equal subintervals, and

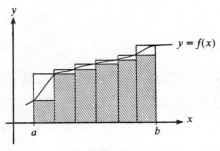

Figure 2.

then constructs an inscribed polygon with (shaded) area A_n and a circumscribed polygon with area $\overline{A}_n$. If $\lim_{n \to \infty}\{\underline{A}_n\} = \lim_{n \to \infty}\{\overline{A}_n\}$, this common value [denoted $\int_a^b f(x)\,dx$] is the desired area sought.

The Definite Integral as Length

The sequence of steps most often employed by elementary calculus texts in determining the length of a "smooth" arc given by a continuous function $y = f(x)$ from $x = a$ to $x = b$ is as follows (Figure 3):

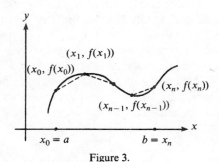

Figure 3.

(i) Divide $[a, b]$ into n equal parts by points $a = x_0 < x_1 < \cdots < x_{n-1} < x_n = b$, and join the consecutive points $(x_i, f(x_i))$ proceeding in order from $i = 0$ to $i = n$. The length of this piecewise linear path then approximates, and is bounded above by, the arc length we are attempting to evaluate.

(ii) As n increases to infinity, the sequence of corresponding piecewise linear path lengths is nondecreasing, bounded above by the arc length that is to be evaluated, and appears to give a better and better approximation of the desired arc length. Therefore, "as in the case of evaluating plane areas," the arc length is *defined* as the limit of this sequence of piecewise linear path lengths. After expressing this limit in terms of a definite integral, we finally obtain the well-known formula for arc length:

$$\text{Arc Length} = \int_a^b \sqrt{1 + \{f'(x)\}^2}\ dx.$$

A Comparison of Plane Areas and Arc Length

The error or gap referred to in the introduction of this article concerns step (ii) in the evaluation of arc length. Specifically, we are referring to the phrase "as in the case of evaluating plane areas" or whatever variation of this phrase is employed in the particular presentation. The point is that step (ii) for arc length is not at all similar to the arguments for evaluating plane areas, and so it cannot depend on the latter for justification. For one thing, while the inscribed rectangles actually "use up" the area to be evaluated in Figure 2, the piecewise linear paths of Figure 3 do not "use up" any part of the arc that it is approximating except at a countable number of points. (We omit the trivial case where the arc itself is a piecewise linear path.) Instead of "exhausting" the point set comprising the arc, the approximating piecewise linear path simply comes physically close to it. In other words, this method of evaluating arc length cannot, as was the case for evaluating planar areas, be justified by the intuitively clear "method of exhaustion" since it is really not exhaustive at all.

How then is this definition of arc length as the limit of the corresponding sequence of piecewise linear path lengths to be justified? One way is by appealing to the fact that as n (the number of linear segments in the approximating path) increases, each of the individual linear segments appears to give a better and better approximation of its part of the arc. But this argument is spurious since physical proximity does not necessarily imply closeness of approximation of length. For example, the true length of the diagonal (Figure 4) is $\sqrt{2}$, but the sum of the lengths of the approximating polygonal paths is always 2.

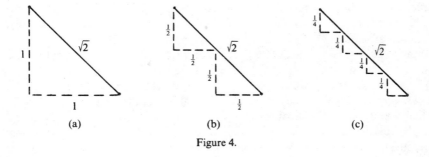

Figure 4.

If we try to use the sandwiching procedure for arc length, we immediately run into another problem. While the sequence of inscribed piecewise linear path lengths forms a sequence of lower bounds for the arc length to be evaluated (the shortest distance between two points is the straight-line segment joining these points), there does not appear in general any way to obtain a suitable corresponding sequence of upper bounds. Therefore, in general, there is no way to define a sequence of "circumscribed piecewise linear paths" so that the method of sandwiching can be employed.

We are therefore left with the following incomplete justification for the arc length formula:

(1) For any positive integer n, the length of the inscribed piecewise-linear path is bounded above by the arc length that is to be evaluated.

(2) The sequence of inscribed piecewise-linear path lengths is monotonically nondecreasing as n increases to infinity.

Statement (1) asserts that the arc length we are seeking to evaluate is an upper bound for the sequence of approximating path lengths while, by statement (2), the value given by the formula

$$L = \int_a^b \sqrt{1 + \{f'(x)\}^2} \, dx$$

is the least upper bound for the set of approximating path lengths. (This is because the limit of a monotonic nondecreasing sequence is equal to the least upper bound of the set of values of the sequence.) But how do we know that the true arc length really is this least upper bound? This gap is filled in the case of plane areas by either "exhaustion" or "sandwiching" arguments. Since neither of these arguments applies to the arc length situation, the gap remains and the veracity of the arc length formula is, of necessity, incomplete.

Conclusion

The point raised in this article also demonstrates that the same philosophical incompleteness exists in other applications of the definite integral. For example, the justification for using the definite integral to evaluate a volume is complete since the method employed is to "exhaust" the volume to be evaluated by inscribed discs or shells. The evaluation of surface areas in three-dimensional space, however, uses physical approximation by two-dimensional areas rather than either "exhaustion" or "sandwiching" and so its justification is incomplete. In general, incomplete justification of this type occurs whenever we try to evaluate a k-dimensional measurement in a *higher dimensional n*-space (that is, $n > k$). This was illustrated in this article when we attempted to evaluate arc length (a 1-dimensional measurement) in the plane (2-dimensional space) and surface area (a 2-dimensional measurement) in space (3-dimensional space). When the dimensions of the measurement being made and the space in which the measurement is being made are the same (such as measuring area in the plane or volume in 3-dimensional space) no incompleteness occurs.

The Formula for Arc Length Does Measure Arc Length

Warren Page, *New York City Technical College, Brooklyn, NY*

Kenneth Goldberg is correct! There is indeed a logical gap in the standard development of the formula for arc length—a gap that appears to have eluded a sizable portion of the mathematics community.

One approach for removing this gap may be found in [2]; other approaches must also exist. However, the following approach, based on a note by Coolidge [1], is noteworthy for three reasons:

(a) It is geometrical in nature, and therefore in the same spirit as the initial development of the formula for arc length.

(b) It uses the method of "sandwiching," thereby corroborating the fact that one-dimensional and two-dimensional geometry are not so intrinsically asymmetrical after all.

(c) it demonstrates how historical perspectives can play an important role in helping students to enhance their understanding and appreciation of the development of mathematical concepts.

1. Length of Arc Defined. Let C be a "smooth" curve given by a continuously differentiable function $y = f(x)$ from $x = a$ to $x = b$. Divide

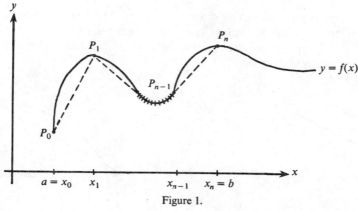

Figure 1.

111

$[a, b]$ into n equal parts by the points $a = x_0 < x_1 < \cdots < x_{n-1} < x_n = b$, and let $P_i = (x_i, f(x_i))$ for $i = 0, 1, \ldots, n$.

The length l_n of the polygonal path $P_0 P_1 \ldots P_n$ clearly approximates, and is bounded above by, the arc length L we are attempting to evaluate. As n increases to infinity, the sequence of polygonal paths gives a better and better visual approximation to C, and so L is defined to be the limit of this nondecreasing sequence $\{l_n\}$:

$$L \stackrel{\Delta}{=} \lim_n l_n. \tag{1}$$

Since f' is continuous (C is "smooth"), one can express $\lim_n l_n$ as a limit of Riemann sums in order to obtain the well-known formula

$$L = \int_a^b \sqrt{1 + \{f'(x)\}^2} \, dx. \tag{2}$$

The critical issue, raised by Goldberg, is the apparent lack of justification for defining L as in (1). *Although $l_n \leqslant L$ for every n, how do we know that $\lim_n l_n$ actually equals L?* And one can also ask: *What does the right-hand side of (2) really measure?*

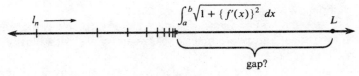

Figure 2.

2. Justification for the Definition of Arc Length.

The earliest geometers had great difficulty comparing and calculating lengths of curves. It was Archimedes who initiated the first great steps toward resolving these difficulties. In his *Sphere and Cylinder*, he begins with the definition:

A curve is *concave in one direction* if every line which connects two of its points has all of the curve (between the two joined points) lying on one side of the line or on the line itself and none on the other side.

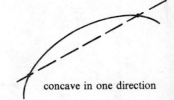

not concave in one direction   concave in one direction

Figure 3.

It is then stated* that:

(A) Of all lines connecting two points, the straight line is the shortest.

(B) Of all lines in the plane having the same extremities, two are unequal when both are concave in the same direction, and that one of them included between the other and the straight line connecting the two points is the lesser.

(C) The length of a curve is a magnitude as defined by Euclid X.

While (A) and (C) are always presumed in the development of the formula for arc length, (B) appears to be unknown or carelessly overlooked. Here we shall see the pivotal role that (B) plays in allowing us to use the method of "sandwiching" to justify defining L as in (1).

Assume that a curve, having length L, is partitioned into n subsections by the set of points $\pi = \{P_0, P_1, \ldots, P_n\}$. Let T_i $(i = 0, 1, 2, \ldots, n)$ denote the tangent line to the curve at P_i, and let Q_i be the point where T_i and T_{i+1} intersect. The projection of Q_i onto chord $P_i P_{i+1}$ will be denoted M_i.

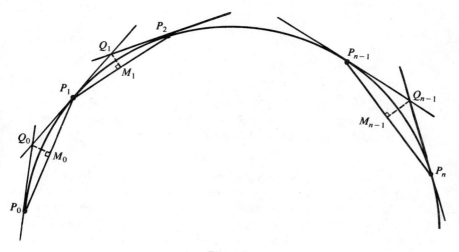

Figure 4.

Let l_π denote the length of the inscribed polygonal path $P_0 P_1 \ldots P_n$, and let $\mathcal{L}_\pi$ denote the length of the circumscribed polygonal path $P_0 Q_0 P_1 Q_1 \ldots P_{n-1} Q_{n-1} P_n$. By virtue of (B), we have

$$l_\pi \leqslant L \leqslant \mathcal{L}_\pi \quad \text{and} \quad 1 \leqslant \frac{L}{l_\pi} \leqslant \frac{\mathcal{L}_\pi}{l_\pi}. \tag{3}$$

---

*Statements (A) and (C) are axioms; statement (B) may be viewed as a definition of order of magnitude between curves that are concave in the same direction.

Now observe that

$$\frac{P_0 Q_0}{P_0 M_0} = \sec \sphericalangle Q_0 P_0 M_0 \quad \text{and} \quad \frac{Q_0 P_1}{M_0 P_1} = \sec \sphericalangle Q_0 P_1 P_0,$$

from which it follows that both ratios approach unity as the length $|P_0 P_1|$ of chord $P_0 P_1$ approaches zero. Thus,

$$\frac{P_0 Q_0 + Q_0 P_1}{P_0 P_1} = \frac{P_0 Q_0 + Q_0 P_1}{P_0 M_0 + M_0 P_1} \to 1 \quad \text{as} \quad |P_0 P_1| \to 0. \qquad (4)$$

In similar fashion,

$$\frac{P_i Q_i + Q_i P_{i+1}}{P_i P_{i+1}} \to 1 \quad \text{as} \quad |P_i P_{i+1}| \to 0 \qquad (5)$$

for each $i = 1, 2, \ldots, n - 1$. Let

$$\|\pi\| = \max_{0 \leqslant i \leqslant n-1} |P_i P_{i+1}|.$$

Then, by (4) and (5), we have

$$\frac{\mathcal{L}_\pi}{l_\pi} \to 1 \quad \text{as} \quad \|\pi\| \to 0.$$

This, in conjunction with (3), establishes that

$$l_\pi \to L \quad \text{as} \quad \|\pi\| \to 0. \qquad (6)$$

It should be clear that (6) holds for every partition $\pi = \{P_0, P_1, \ldots, P_n\}$. In particular, consider $P_i = (x_i, f(x_i))$ as in section 1. Notation-wise, l_π became l_n. Since $\|\pi\| \to 0$ as $n \to \infty$, we have

$$l_n \to L \quad \text{as} \quad n \to \infty.$$

Thus, there is no gap as conjectured in Figure 2, and (1) is indeed correct for curves that are concave in one direction. Since every "smooth" curve can be broken into small subsections, each of which is concave in one direction, the veracity of (1) for arbitrary smooth curves follows. Accordingly, the formula for arc length does measure arc length.

For a more historical account of this discussion, and related matters concerning lengths of curves, see [1].

REFERENCES
1. J. L. Coolidge, The lengths of curves, Amer. Math. Monthly, 60 (1953) 89–93.
2. Gasper Goffman, Arc length, Amer. Math. Monthly, 71 (1964) 303–304.

Rational Solutions of $x^y = y^x$

Charles S. Weaver, *University of Illinois, Urbana, IL*

The function x^y is often considered in elementary calculus (see [2], [3] for example). In particular, one asks whether the equation $x^y = y^x$ has solutions for $x \neq y$. This paper shows how a search for solutions leads from calculus to number theory, and eventually to a representation of all solutions of $x^y = y^x$.

An analytic approach to the solution begins by taking natural logarithms of $x^y = y^x$ to obtain $x/\ln x = y/\ln y$. Elementary calculus can then be used to study the function $f(t) = t/\ln t$. The graph of $f(t)$ is u-shaped for $t > 1$, has a minimum point at (e, e), and approaches infinity as t becomes infinite. Furthermore, $\lim_{t \to 1^-} f(t) = -\infty$ and $\lim_{t \to 1^+} f(t) = \infty$. The graph shows that $f(t)$ assumes every value greater than e twice—once at a point $x \in (1, e)$, and again at some point $y > e$. Thus for every $x \in (1, e)$, there is a $y > e$ such that $x^y = y^x$.

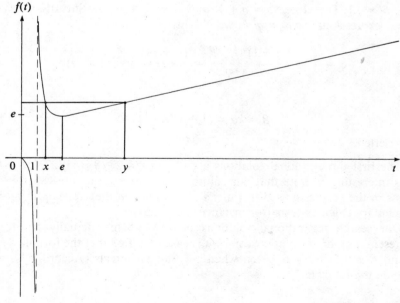

There are infinitely many pairs (x, y) satisfying $x^y = y^x$ [in fact, one pair for each $x \in (1, e)$]. One such pair is $(2, 4)$. Actually $(2, 4)$ is the only integral solution of $x^y = y^x$ since 2 is the only integer in $(1, e)$. In contrast to this, there exists a countably infinite collection of rational pairs (i.e., pairs of rational numbers) solving $x^y = y^x$. The principal result is the following.

Theorem. *A rational pair (α, β) with $1 < \alpha < \beta$ satisfies $\alpha^\beta = \beta^\alpha$ if and only if $\alpha = (1 + 1/q)^q$ and $\beta = (1 + 1/q)^{q+1}$ for some positive integer q.*

Proof. Since every rational pair (α, β) as above satisfies $\alpha^\beta = \beta^\alpha$, it remains to prove that only numbers of this form work. Therefore, assume that (α, β) with $1 < \alpha < \beta$ satisfies $\alpha^\beta = \beta^\alpha$. Let $\beta = \alpha^k$ and let $\alpha = m/n$ for positive, relatively prime integers m, n. Then $\alpha^\beta = (\alpha^k)^\alpha$ gives $k = \beta/\alpha > 1$ and $k = \alpha^k/\alpha = \alpha^{k-1}$. Since k is rational, assume that $k = p/q$ for positive relatively prime integers p, q. Using $\alpha = m/n$ and $k = p/q$ in $k = \alpha^{k-1}$, we have

$$\frac{m}{n} = \left(\frac{p}{q} \right)^{q/(p-q)} = \frac{p^{q/(p-q)}}{q^{q/(p-q)}} .$$

Since m, n (and p, q) are relatively prime, numerators and denominators can be equated to show that $p^{q/(p-q)} = m$ and $q^{q/(p-q)} = n$ are integers. Since q and $p - q$ are relatively prime, $p^{1/(p-q)} = a$ and $q^{1/(p-q)} = b$ are integers. Let $j = p - q$ so that $j = a^j - b^j$. This requires that $j = 1$. [Otherwise, $a = b + c$ for some integer $c > 0$ and $a^j = (b + c)^j > b^j + j$ contradicts $a^j - b^j = j$.] Therefore, $p = q + 1$ and $k = (q + 1)/q$. Substituting the above expressions for m, n, p, k, we obtain

$$\alpha = \frac{m}{n} = \frac{(q + 1)^{q/(p-q)}}{q^{q/(p-q)}} = \frac{(q + 1)^q}{q^q} = (1 + 1/q)^q$$

and

$$\beta = k\alpha = (1 + 1/q)^{q+1}$$

as asserted.

The first three rational solutions are $(2, 4), (9/4, 27/8), (64/27, 256/81)$. It is interesting to note that our solution pairs to $x^y = y^x$ are exactly the terms in the sequence used in Euler's definition of e that appears in many elementary calculus texts (for instance [1], p. 267).

Our search for rational solutions of $x^y = y^x$ has actually led to a representation of all solutions, since it is easily verified that the formulas for α and β satisfy $\alpha^\beta = \beta^\alpha$ even when q is not an integer. Accordingly, we include the following.

Corollary. *All solutions of $x^y = y^x$ are of the form*

$$x = (1 + 1/t)^t \quad and \quad y = (1 + 1/t)^{t+1} \qquad (t > 0).$$

Ed. Note. For a comprehensive development of this topic, see R. Arthur Knoebel's "Exponentials Reiterated," *Amer. Math. Monthly*, 88 (1981) 235–252.

REFERENCES

1. R. E. Johnson and F. L. Kiokemeister, Calculus with Analytic Geometry, 2nd ed., Allyn and Bacon, Boston, 1960.
2. T. W. Shilgalis, Graphical solution of the equation $a^b = b^a$, Math. Teacher, 66 (March 1973) 235.
3. J. T. Varner, III, Comparing a^b and b^a using elementary calculus, TYCMJ, 7 (December 1976) 46.

A Vectorial Approach to "The Chase"

Sidney H. L. Kung, *Jacksonville University, Jacksonville, FL*

In this note we shall solve an old pursuit problem ([2], [4]) by using vector calculus. Our problem states:

> *A cat and mouse, D_0 distance apart, suddenly spot each other. The mouse begins to run along a straight path at constant speed v_m, and the cat always runs directly toward the mouse with constant speed $v_c > v_m$. How long will it take for the cat to catch the mouse?*

Our method of answering this question may not be new, but no earlier references ([1], [3], [5], [6], [7]) have indicated this approach.

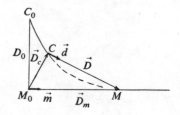

Let $\vec{D}_c$ and $\vec{D}_m$ be the position vectors of the cat (C) and the mouse (M). Assume that $\vec{m}$ is a unit vector along the mouse's path, and let $\vec{d}$ be a unit vector directed from C to M. The relative position vector $\vec{D} = \vec{D}_m - \vec{D}_c$ will be written $\vec{D} = D\vec{d}$, where D is the magnitude of $\vec{D}$. Since $C = C(t)$ and $M = M(t)$, the vectors $\vec{D}_c$, $\vec{D}_m$, $\vec{d}$ (therefore, $\vec{D}$) and the scalar D are all functions of time t. Clearly,

$$\frac{d(\vec{D}_m)}{dt} = v_m\vec{m} \quad \text{and} \quad \frac{d(\vec{D}_c)}{dt} = v_c\vec{d}.$$

Therefore,

$$\frac{d(\vec{D})}{dt} = v_m\vec{m} - v_c\vec{d}. \tag{1}$$

Since $d(\vec{m})/dt = 0$, we can differentiate the equation $\vec{D} \cdot \vec{m} = (D\vec{d} \cdot \vec{m})$ to obtain

$$\frac{d(\vec{D})}{dt} \cdot \vec{m} = \frac{d}{dt}(D\vec{d} \cdot \vec{m})$$

and then substitute (1) to get

$$v_m - (v_c \vec{d} \cdot \vec{m}) = \frac{d}{dt}(D\vec{d} \cdot \vec{m}). \tag{2}$$

It follows, from $\vec{D} \cdot (d(\vec{d})/dt) = 0$ and $\vec{d} \cdot \vec{d} = 1$, that

$$\vec{D} \cdot \frac{d(\vec{D})}{dt} = \vec{D} \cdot \frac{d(D\vec{d})}{dt} = D\frac{d(D)}{dt}.$$

One can also readily verify, using (1), that

$$\vec{D} \cdot \frac{d(\vec{D})}{dt} = (D\vec{d}) \cdot (v_m \vec{m} - v_c \vec{d}) = Dv_m(\vec{d} \cdot \vec{m}) - Dv_c.$$

Therefore the two expressions for $\vec{D} \cdot \dfrac{d(\vec{D})}{dt}$ can be equated, and division by D yields

$$\frac{d(D)}{dt} = v_m \vec{d} \cdot \vec{m} - v_c. \tag{3}$$

Now multiply (2) by v_m and multiply (3) by v_c. Add these results and simplify in order to obtain

$$v_m \frac{d(D\vec{d} \cdot \vec{m})}{dt} + v_c \frac{d(D)}{dt} = v_m^2 - v_c^2. \tag{4}$$

Integrating (4), we get

$$v_m D\vec{d} \cdot \vec{m} + v_c D = (v_m^2 - v_c^2)t + A. \tag{5}$$

At $t = 0$, we have $D = D_0$ and $\vec{D} = \vec{D}_0$. Therefore

$$A = D_0(v_m \vec{D}_0 \cdot \vec{m} + v_c).$$

To find the time of capture, set $D = 0$ and obtain

$$t = \frac{D_0(v_m \vec{D}_0 \cdot \vec{m} + v_c)}{v_c^2 - v_m^2}. \tag{6}$$

If the initial vector $\vec{D}_0$, from cat to mouse, is perpendicular to the mouse's path, then $\vec{D}_0 \cdot \vec{m} = 0$ and (6) reduces to

$$t = \frac{D_0 v_c}{v_c^2 - v_m^2}.$$

REFERENCES

1. R. Bellman, Modern Elementary Differential Equations, Addison-Wesley, Reading, Mass., 1968, pp. 120–124.
2. Pierre Bourguer, Lignes de poursuite, Mémoires de l'Académie Royale des Sciences (1732), 1–14.
3. Gerald Crough, A pursuit problem, Math. Mag., 44 (1971) 94–97.
4. H. A. Luther, Solution to Problem 3942, Amer. Math. Monthly, 48 (1941) 484–485.
5. A. L. Nelson, K. W. Folley, and M. Coral, Differential Equations, D. C. Heath, Boston, Mass., 1964, pp. 162–165.
6. W. F. Osgood, Advanced Calculus, Macmillan, New York, 1937, p. 332.
7. M. Tenenbaum and H. Pollard, Ordinary Differential Equations, Harper and Row, New York, 1963, pp. 178–182.

Discontinuity Amidst Differentiability

Michael W. Ecker, *Pennsylvania State University, Worthington Scranton Campus, Scranton, PA*

What mixture of continuity, discontinuity, and differentiability can a function enjoy? Some interesting illustrations will be given to show that the possibilities can be quite surprising. We invite the reader to further explore these results in the exercises that follow.*

1. Preliminaries. Throughout this paper we are speaking of functions $f: \mathbb{R} \to \mathbb{R}$ mapping the set of reals to the reals. It will be helpful to recall some known results and introduce a few basic ingredients for our later development.

Every function differentiable at $x \in \mathbb{R}$ is also continuous at x. The converse, of course, fails—as the standard example $f(x) = |x|$ at $x = 0$ demonstrates. It is interesting to note ([5], [7], [8]) that there exist continuous, nowhere differentiable functions.

It is known (see [4], for example) that no function can be both continuous on the set $\mathfrak{Q}$ of rationals and nowhere continuous on the set $\mathfrak{I}$ of irrationals. Surprisingly, the reverse is true when these underlying sets are interchanged.

Example 1. The "ruler function"

$$r(x) = \begin{cases} 0, & x \in \mathfrak{I}, \\ \dfrac{1}{q}, & x = \dfrac{p}{q} \text{ in lowest terms,} \end{cases}$$

is continuous on $\mathfrak{I}$ and nowhere continuous on $\mathfrak{Q}$. Furthermore, r is nowhere differentiable on $\mathfrak{I}$ (therefore, nowhere differentiable on $\mathbb{R}$).

Exercise 1. Verify that r has the properties asserted in Example 1.

[*Hint*: If r is discontinuous at $y \in \mathfrak{I}$, there is a sequence (p_i/q_i) of rational numbers, in reduced form, such that $(p_i/q_i) \to y$ and $(1/q_i) \not\to 0$. Since

---

*The author wishes to thank Warren Page for his contributions and assistance in enhancing this article.

$\{q_i : i \in \mathbb{N}\}$ is a finite set of integers, $q_i \leqslant M$ for some number M. Conclude, from $(p_i/q_i) \geqslant p_i/M$, that $p_i \nrightarrow 0$ and that $\{p_i : i \in \mathbb{N}\}$ is finite. Therefore, $\{(p_i/q_i) : i \in \mathbb{N}\}$ is a finite set of rationals, and this contradicts $(p_i/q_i) \rightarrow y$.]

A subset E of a set $F \subset \mathbb{R}$ is said to be *dense in F* if for each $x \in F$ and each $\delta > 0$, there exists an element $y \neq x$ in $(x - \delta, x + \delta) \cap E$. Intuitively, E being dense in F means that each point of F has points of E arbitrarily close to it.

Example 2. The set $E = \cup_{i=0}^{n-1}(i/n, (i + 1)/n)$ is dense in $F = [0, 1]$. As also shown below, $\mathcal{Q}$ and $\mathcal{I}$ are dense in $\mathbb{R}$; the set of natural numbers is not dense in $\mathbb{R}$.

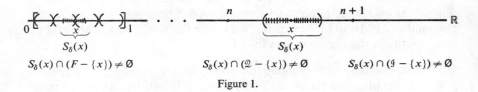

$$S_\delta(x) \cap (F - \{x\}) \neq \varnothing \qquad S_\delta(x) \cap (\mathcal{Q} - \{x\}) \neq \varnothing \qquad S_\delta(x) \cap (\mathcal{I} - \{x\}) \neq \varnothing$$

Figure 1.

Example 3. Every number $x \in (0, 1)$ can be written in decimal form $x = \sum_{n=1}^{\infty} a_n/10^n$ for digits $a_1, a_2, \ldots$. Whenever a choice of representation exists, we write x as an infinite decimal $.a_1 a_2 \ldots a_k \bar{0}$, rather than as either a finite decimal $.a_1 a_2 \ldots a_k$ or an infinite decimal $.a_1 a_2 \ldots a_k \bar{9}$ with a tail of nines. Thus, $1/20$ is written as $.05\bar{0}$, rather than as $.05$ or $.04\bar{9}$.

Let $\mathcal{T}$ denote the set of terminal decimals in $(0, 1)$, and let $\mathfrak{N}\mathcal{T}$ denote the set of nonterminating decimals in $(0, 1)$. Clearly, $\mathcal{T} \subset \mathcal{Q}$ and $\mathcal{I} \subset \mathfrak{N}\mathcal{T}$. Equality does not hold. For instance, $1/3 \in \mathcal{Q} \cap \mathfrak{N}\mathcal{T}$ and $1/3 \notin \mathcal{T} \cup \mathcal{I}$. Let $\mathcal{P}$ denote the rationals in $(0, 1)$ which have an infinite periodic decimal representation (as, for instance, $.05\bar{8}$ and $1/3 = .\bar{3}$). Then

$$\mathcal{T} = \mathcal{Q} - \mathcal{P} \quad \text{and} \quad \mathfrak{N}\mathcal{T} = \mathcal{I} \cup \mathcal{P}.$$

The reader can easily verify that $\mathcal{T}$ and $\mathfrak{N}\mathcal{T}$ are each dense in $(0, 1)$ since every decimal $.a_1 a_2 \ldots a_k a_{k+1} \ldots$ has decimals $.a_1 a_2 \ldots a_k \bar{0} \in \mathcal{T}$ and, say, $.a_1 a_2 \ldots a_k \bar{1} \in \mathcal{P} \subset \mathfrak{N}\mathcal{T}$ "near" it.

2. Permutation-Induced Functions.

We have already seen that $r(x)$ is continuous on one dense set $\mathcal{I} \subset \mathbb{R}$, nowhere continuous on another dense set $Q \subset \mathbb{R}$, and is nowhere differentiable on $\mathbb{R} = \mathcal{I} \cup \mathcal{Q}$. A rather interesting class of functions with equally unusual properties will now be introduced.

As in Example 3, assume that every $x \in (0, 1)$ carries a unique decimal representation, and let $\mathfrak{D} = \{0, 1, \ldots, 9\}$. Then every permutation $p : \mathfrak{D}$

$\rightarrow \mathfrak{D}$ induces a function

$$p^* : (0,1) \rightarrow [0,1]$$

$$x = \sum_{x=1}^{\infty} a_n/10^n \leadsto \sum_{n=1}^{\infty} p(a_n)/10^n.$$

Note that our uniqueness of representation for x guarantees that every mapping $p^*(.a_1 a_2 \ldots a_n \ldots) = .p(a_1)p(a_2) \ldots p(a_n) \ldots$ is well-defined.

Since $\mathfrak{D}$ has ten elements, there are 10! permutation-induced functions p^*. A few examples of such functions are illustrated below.

Example 4.   (a) Let $p_1(k) = k$ and $p_2(k) = 9 - k$ for each $k \in \mathfrak{D}$. Then $p_1^*(x) = x$ and $p_2^*(x) = 1 - x$. The latter follows from

$$p_2^*(x) = \sum_{n=1}^{\infty} p_2(a_n)/10^n = \sum_{n=1}^{\infty} (9 - a_n)/10^n = \sum_{n=1}^{\infty} 9/10^n - \sum_{n=1}^{\infty} a_n/10^n = 1 - x.$$

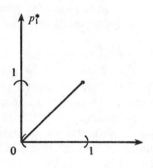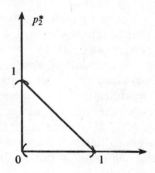

Figure 2.

(b) Let $p_3(k) = k + (-1)^k$ for $k \in \mathfrak{D}$. Observe, for example, that $p_3^*(1/3)$ $= p_3^*(.\overline{3}) = .\overline{2}$ and $p_3^*(.29\overline{0}) = .38\overline{1}$. It will follow, from Property 2 of this section, that p_3^* is nowhere continuous on $\mathfrak{I}$. To get some feeling as to why this is so, consider the continuity of p_3^* at $x = .2590 \in \mathfrak{I}$. Continuity requires that $p_3^*(y)$ can be made arbitrarily close to $p_3^*(x) = .348\overline{1}$ by taking y sufficiently near to x. This, however, is impossible for $y < x$. For instance,

$$\begin{aligned} x &= .259000 \ldots \\ y &= .2589999 \ldots a_m \ldots \end{aligned} \Rightarrow \begin{aligned} p_3^*(x) &= .348111 \ldots \\ p_3^*(y) &= .3498888 \ldots b_m \ldots \end{aligned}$$

shows that $p_3^*(y) - p_3^*(x) > .001$. Thus, $p_3^*(y)$ cannot be made arbitrarily close to $p_3^*(x)$ no matter how near y is taken, from below, to x.

(c) The "breakdown" in continuity of p_3^* in (b) would not have materialized if we had considered only $y > x$. In the language of limits, p_3^* is *continuous from above at* x. That is, $p_3^*(x)$ is defined and $\lim_{y \to x^+} p_3^*(y)$ $= p_3^*(x)$. Every p^* is continuous from above at each $x \in (0, 1)$. The reader can easily verify this by noting that any $x = .a_1 a_2 \ldots a_k a_{k+1} \ldots$ and $y = .b_1 b_2 \ldots b_k b_{k+1} \ldots$ with $b_{k+1} > a_{k+1}$ (permissible since infinitely many a_j must be less than 9) have images within 10^{-k} of each other.

Some key characteristics of these p^* functions are outlined below (fuller details appear in [1], [2], [3]). Our starting point is the observation that

$$(0, 1) = \mathfrak{I} \cup \mathfrak{N}\mathfrak{I} \quad \text{and} \quad \mathfrak{I} \cap \mathfrak{N}\mathfrak{I} = \emptyset.$$

Property 1. *Every p^* is continuous on $\mathfrak{N}\mathfrak{I}$.*

Proof. Consider any $x \in \mathfrak{N}\mathfrak{I}$, and let $\epsilon > 0$ be given. Then $1/10^k < \epsilon$ for some natural number k. Since $x \in \mathfrak{N}\mathfrak{I}$, its decimal representation cannot be of the form $.a_1 a_2 \ldots a_k a_{k+1} \ldots \bar{b}$ for $b = 0$ or $b = 9$. Let i be the first index after k for which $a_i \neq 0$, and let j be the first index after k for which $a_j \neq 9$. For any $y \in (0, 1)$ satisfying $|y - x| < 1/10^{i+j}$, it is clear that y and x have the same beginning k digits in their decimal representations, and so $|p^*(y) - p^*(x)| < 10^{-k} < \epsilon$.

The polynomials p_1^* and p_2^* are obviously differentiable (therefore continuous) on $(0, 1)$. But what about the remaining $10! - 2$ such p^* functions? By a modification of the proof of property 2, below, it can be shown that every p^*, other than p_1^* and p_2^*, is nowhere continuous on (at least) a dense subset of $\mathfrak{I}$. (For more insight, see [2].) For our purposes, we shall restrict ourselves to those p^* that are discontinuous on *all* of $\mathfrak{I}$. As will be made clear in the proof to follow, this involves permutations p such that $\{p(0), p(9)\} \neq \{0, 9\}$. Let p_e denote any such p satisfying this restriction.

Property 2. *Every p_e^* is nowhere continuous on $\mathfrak{I}$.*

Proof. We shall show that if p^* is continuous at some $x \in \mathfrak{I}$, then p^* cannot be any p_e^*. The proof of Property 2 will then follow by contraposition. Since every p^* is continuous from above (Example 4c), it suffices to assume that p^* is continuous from below at some $x = .a_1 a_2 \ldots a_k \bar{0} \in \mathfrak{I}$. Let $\epsilon \in (0, 1/10^{k+1})$ and choose $y \in (0, 1)$ to be of the form $y = .a_1 a_2 \ldots a_{k-1} b_k 999 \ldots 9 b_{k+m} \bar{0}$, where $b_k = a_k - 1$ and $b_{k+m} = 9$. Observe that $0 < x - y = 1/10^{k+m}$. Clearly, y can be made arbitrarily near to x by taking sufficiently many nines in the tail $b_{k+1} \ldots b_{k+m}$. Now consider $p^*(x) - p^*(y)$:

$$
\begin{array}{lllll}
p^*(x) = & .p(a_1)p(a_2) \ldots p(a_{k-1}) & \mid p(a_k)p(0)p(0) \ldots p(0) & \mid p(0) \ldots \\
p^*(y) = & .p(a_1)p(a_2) \ldots p(a_{k-1}) & \mid p(b_k)p(9)p(9) \ldots p(9) & \mid p(0) \ldots \\
\hline
p^*(x) - p^*(y) = & . & \mid & \mid \bar{0}
\end{array}
$$

If $p^*(x) > p^*(y)$, then $p^*(x) - p^*(y) < 1/10^{k+1}$ requires that $p(b_k)$ $= p(a_k) - 1$ and $p(0) < p(9)$. Therefore (after the first subtraction), $[10 + p(0)] - [p(9) + 1] = 0$, and so $p(0) + 9 = p(9)$. Since $p(0), p(9) \in \mathfrak{D}$, this yields $p(0) = 0$ and $p(9) = 9$.

If $p^*(x) < p^*(y)$, then $p^*(y) - p^*(x) < 1/10^{k+1}$ requires that $p(a_k)$ $= p(b_k) - 1$ and $p(9) < p(0)$. Therefore, $[10 + p(9)] - [p(0) + 1] = 0$ yields $p(9) + 9 = p(0)$, and this requires that $p(9) = 0$ and $p(0) = 9$.

The continuity (from below) of p^* at $x \in \mathfrak{T}$ means that $|p^*(x) - p^*(y)|$ $< \epsilon$ for sufficiently small $x - y > 0$. And, as we have seen, this necessitates that $\{p(0), p(9)\} \neq \{0, 9\}$. Thus, p^* cannot be any p_e^* mapping.

To complete the picture, we also prove:

Property 3. *All p^*, other than p_1^* and p_2^*, are nowhere differentiable on* $(0, 1)$.

Proof. Suppose that an arbitrary p^* is differentiable at any one $x = .a_1 a_2 \ldots a_n \ldots$ in $(0, 1)$. We will show that this implies that either $p = p_1$ or $p = p_2$. The result follows by contraposition.

Now, some decimal digit $d \neq 9$ must occur infinitely often amongst the digits a_n of x. Otherwise, this representation of x would terminate in nines and thus be impermissible. (For instance, if $x \in \mathfrak{T}$, then $d = 0$.) Let $\{n_m : m = 1, 2, \ldots\}$ denote the subsequence of indices on a_n for which $a_n = d$. Also let $h_m = s/10^{n_m}$, where s is any of the nine nonzero values in the progression $\{0 - d, 1 - d, \ldots, 9 - d\}$. Note that the decimal representations of x and $x + h_m$ will differ only in the n_mth digit.

Now consider the difference quotient

$$\frac{p^*(x + h_m) - p^*(x)}{h_m}$$

$$= \frac{1}{s/10^{n_m}} \Big[.p(a_1)p(a_2) \ldots p(a_{n_m-1})p(s + d)p(a_{n_m+1}) \ldots$$

$$- .p(a_1)p(a_2) \ldots p(a_{n_m-1})p(d)p(a_{n_m+1}) \ldots \Big]$$

$$= \frac{\{p(s + d) - p(d)\} \cdot 10^{-n_m}}{s \cdot 10^{-n_m}}.$$

After simplifying, we see that

$$\frac{p^*(x + h_m) - p^*(x)}{h_m} = \frac{p(s + d) - p(d)}{s}$$

for each $m = 1, 2, \ldots$. Since p^* is assumed to be differentiable at x, its derivative at x is some constant value c—and this c is independent of $s \neq 0$ in $\{0 - d, 1 - d, \ldots, 9 - d\}$. Thus,

$$\frac{p(s + d) - p(d)}{s} = c.$$

Since $s \neq 0$ implies that $s + d \neq d$, it is clear that $p(s + d) \neq p(d)$. Thus, $c \neq 0$. Taking $s = 1$, one can also see that $c = p(1 + d) - p(d)$ is an integer between ± 9. Actually, $c = \pm 1$. To see this, note that

$$p(s + d) - p(d) = cs \tag{1}$$

and that the numbers $p(s + d) - p(d)$ are some reordering of the nine nonzero values among the set $\{0 - p(d), 1 - p(d), \ldots, 9 - p(d)\}$ which, in turn, consists of nine almost consecutive integers that include either $\{1, 2, 3, 4, 5\}$ or their negatives. By (1), c is a division of these, and the only integers $c \in [-9, 9]$ with this property are ± 1.

We now have, by taking $c = \pm 1$ in (1),

$$p(s + d) - p(d) = \pm s \tag{2}$$

for each nonzero $s \in \{0 - d, 1 - d, \ldots, 9 - d\}$. Equation (2), of course, is also true for $s = 0$. Summing (2), we obtain

$$\sum_{s=0-d}^{9-d} p(s + d) - p(d) = \pm \sum_{s=0-d}^{9-d} s$$

and so

$$\sum_{s=0-d}^{9-d} p(s + d) - 10p(d) = \pm 5(9 - 2d). \tag{3}$$

Since $\{p(s + d) : s = 0 - d, 1 - d, \ldots, 9 - d\} = \mathcal{D}$, we have

$$\sum_{s=0-d}^{9-d} p(s + d) = \sum_{i=0}^{9} i = 45.$$

Substituting this into (3), we obtain

$$45 - 10p(d) = \pm 5(9 - 2d)$$

from which it follows that either

$$p(d) = d \quad \text{or} \quad p(d) = 9 - d. \tag{4}$$

The implications of (2) and (4) are twofold: First, $p(s + d) - p(d) = s$ implies that $p(d) = d$. And this yields $p(s + d) - d = s$, or $p(s + d) = s + d$, for each $s + d \in \{0, 1, \ldots, d - 1, d + 1, \ldots, 9\}$. Together, this establishes that $p(k) = k$ for each $k \in \mathcal{D}$. Thus, $p^* = p_1^*$. The second implication arises from $p(s + d) - p(d) = -s$ and its result that $p(d) = 9 - d$. In particular, the preceding two equations yield $p(s + d) - (9 - d) = -s$ for each $s + d \in \{0, 1, \ldots, d - 1, d + 1, \ldots, 9\}$. Thus, $p(d) = 9 - d$ and $p(s + d) = 9 - (s + d)$ for each $s + d \in \{0, 1, \ldots, d - 1, d + 1, \ldots, 9\}$. This means that $p(k) = 9 - k$ for each $k \in \mathcal{D}$, and so $p^* = p_2^*$.

Exercise 2. Show that each function p^* satisfies: $p^*(x) \in \mathcal{Q}$ if and only if $x \in \mathcal{Q}$.

[*Hint*: Use the characterization of the rationals as those reals having repeating decimals.]

Exercise 3. Some p^* are injective (1-1), surjective (onto), or bijective (1 − 1 and onto). The ones that aren't come tantalizingly close. Show that p^* is injective (p^* is surjective) if and only if $p(9) \in \{0, 9\}$. (Hence, this condition is necessary and sufficient for the bijectivity of p^*.) Use this to show that $2/10 = 20\%$ of the 10! functions p^* are bijective.

Exercise 4. Suppose we allowed $x \in \mathfrak{T}$ to be represented with an infinite tail of nines instead of as a terminating decimal (e.g., $\frac{1}{2} = .4\bar{9}$, and not $.5\bar{0}$). Show, by mimicking the earlier proofs, that the major results would be unchanged—except that "continuity from above" would be replaced by "continuity from below." (This indicates the nonremovable nature of the discontinuity.)

Exercise 5. Repeat Exercise 4, assuming that all $x \in \mathfrak{T}$ have only finite decimal representations (e.g., $\frac{1}{2} = .5$).

3. Dense Differentiability.

As the following table indicates, we have already exhibited functions that are continuous on one dense set A, nowhere continuous on another dense set B, and are nowhere differentiable on $A \cup B$.

Function	Continuous on dense A	Nowhere Continuous on dense B	Differentiable on $A \cup B$
$r(x)$	$A = \mathfrak{I}$	$B = \mathfrak{Q}$	nowhere
p_e^*	$A = \mathfrak{R}\mathfrak{T}$	$B = \mathfrak{T}$	nowhere
$g(x) = x^2 \cdot r(x)$	$A = \mathfrak{I} \cup \{0\}$	$B = \mathfrak{Q} - \{0\}$	at $x = 0$ with $g'(0) = 0$
f	A	B	$C \subset A$

The third line of the table shows that a function g, having "dense continuity" and "dense discontinuity," may be differentiable at some $x \in A$. The intriguing question is: Can we extend this table by finding a function which, in addition to dense continuity and dense discontinuity, also has "dense differentiability" (i.e., differentiable on a dense set $C \subset A$)? The answer, as we shall see shortly, is affirmative!

Let $\mathbb{S} = \{x \in \mathfrak{R}\mathfrak{T} : x$ has at most finitely many zeros or nines in its decimal representation}. The first thing to verify is:

Claim 1. $\mathbb{S}$ is dense in $(0, 1)$.

Proof. Consider any $x = \sum_{n=1}^{\infty} a_n / 10^n$ in $(0, 1)$, and let $\delta > 0$ be given. Then $1/10^j < \delta$ for some natural number j. Now consider $y =$

$\sum_{n=1}^{\infty} b_n/10^n$, where

$$b_n = \begin{cases} a_n, & 1 \leqslant n \leqslant j, \\ 1, & j+1 \leqslant n. \end{cases}$$

Evidently, $y \in S$ and $|y - x| \leqslant 1/10^j < \delta$.

For each $x \in \mathcal{T}$, let k denote the largest value of n for which $a_n \neq 0$ (i.e., k is the index of the last nonzero digit in x's expansion). For example, $k = 4$ for .3002. Using this notation, we define the function

$$f : (0, 1) \to [0, 1)$$

$$x \rightsquigarrow \begin{cases} 0, & x \in \mathcal{N}\mathcal{T}, \\ \dfrac{1}{10^{2k+4}}, & x \in \mathcal{T}. \end{cases}$$

For example, $f(\pi/4) = 0$. And $f(.1) = 1/10^6$, $f(.12) = 1/10^8$, $f(.\overline{12}) = 0$. Observe that "larger" terminating decimals have smaller positive images—a fact that will be useful later. Our objective now is to establish:

Claim 2. *f is continuous on $\mathcal{N}\mathcal{T}$ and nowhere continuous on $\mathcal{T}$.*

Proof. For continuity, observe that $x = .a_1 a_2 \ldots a_n \ldots$ has infinitely many $a_n \neq 0, 9$ because $x \in \mathcal{N}\mathcal{T}$, and our convention forbids x from having an infinite tail of nines. Given $\epsilon > 0$, we can therefore find k such that $a_k \neq 0, 9$ and such that $1/10^{2k+4} < \epsilon$. Choose $\delta < 1/10^k$. If $|y - x| < \delta$, then x and y match in at least the first k digits, and $a_k \neq 0$ ensures that $f(y) \leqslant 1/10^{2k+4}$. Thus, $|f(y) - f(x)| = f(y) < \epsilon$.

For the discontinuity assertion, consider any $x = .a_1 a_2 \ldots a_k \overline{0} \in \mathcal{T}$, and let $\epsilon < 1/10^{2k+4}$. Given any $\delta > 0$, choose j large enough so that $1/10^j < \delta$. Then $|y - x| < \delta$ for $y = .a_1 a_2 \ldots a_k \underbrace{000 \ldots 0\overline{1}}_{j}$, say. Since $y \in \mathcal{N}\mathcal{T}$, we also have $|f(y) - f(x)| = |f(x)| > \epsilon$.

Our final proof establishes that f is differentiable on S. In fact:

Claim 3. *$f' = 0$ on S.*

Proof. Since $S \subset \mathcal{N}\mathcal{T}$, the difference quotient for f with $x = \sum_{n=1}^{\infty} a_n/10^n$ in S is

$$\frac{f(x+h) - f(x)}{h} = \frac{f(x+h)}{h} \qquad (h \neq 0).$$

If $x + h \in \mathcal{N}\mathcal{T}$, then $f(x+h) = 0$. Hence, it only remains to show that $f(x+h)/h \to 0$ as $h \to 0$ for $x + h \in \mathcal{T}$.

Given any $\epsilon > 0$, choose a natural number j such that

$$\frac{1}{10^j} < \epsilon \quad \text{and} \quad a_n \notin \{0, 9\} \text{ for } n > j - 2.$$

(The condition on a_n is possible since $x \in \mathbb{S}$.) Now consider $|h| < 1/10^j$. Clearly, h has at least j zeros after the decimal point. Assume that h has exactly m ($m \geqslant j$) leading zeros. Then

$$\frac{1}{10^{m+1}} \leqslant |h| < \frac{1}{10^m}, \tag{5}$$

and one can also visualize the term $x + h$:

$$
\begin{array}{ll}
x = & .a_1 a_2 \cdots a_{j-2} a_{j-1} a_j \cdots a_{m-1} a_m a_{m+1} a_{m+2} \cdots \\
h = & \pm .0\,0 \cdots 0 \quad 0 \quad 0 \cdots 0 \quad 0\, b_{m+1} b_{m+2} \cdots \\
\hline
x+h = & .a_1 a_2 \cdots a_{j-2} a_{j-1} a_j \cdots a_{m-1} c_m c_{m+1} c_{m+2} \cdots
\end{array}
$$

Here we have implicitly used the fact that $a_m \notin \{0,9\}$. If $h > 0$, then $c_m \neq 0$.

If $h < 0$, then either $c_m \neq 0$ or $c_m = 0 \neq a_{m-1}$. Therefore, in either case, $x + h \in \mathbb{T}$ contains at least $m - 1$ decimal digits before the $\bar{0}$ and it follows, using (5), that

$$\left| \frac{f(x+h)}{h} \right| \leqslant \frac{1/10^{2(m-1)+4}}{1/10^{m+1}} = \frac{1}{10^{m+1}} < \frac{1}{10^j} < \epsilon.$$

This establishes that $f'(x) = 0$ for each $x \in \mathbb{S}$.

4. Additional Results.

Some modifications of the ideas in the previous sections can be used to illustrate other unusual "function pathologies."

(a) *More on Dense Differentiability.* The function f in section 3 can be used to prove the following: *Given any differentiable function $F(x)$, there exists a function $g(x)$ which is continuous on a dense subset of its domain, nowhere continuous on another dense subset, and differentiable on some dense subset $\mathbb{S}$ of its domain. Moreover, $g' = F'$ at points where g is differentiable, that is, g' is the restriction of F' to $\mathbb{S}$.*

Proof. Let $g(x) = F(x) + f(x)$, where $f(x)$ is the function of Section 3. By standard calculus theorems on sums and differences of continuous functions and differentiable functions, it is clear that $g(x)$ has precisely the same continuity and differentiability properties as $f(x)$. Thus, $\mathbb{S}$ here is the same set as that used in Section 3, and hence is independent of F. Moreover, $g'(x) = F'(x) + f'(x) = F'(x) + 0 = F'(x)$ on $\mathbb{S}$.

(b) *Another Family of Permutation-Induced Functions.* In Section 2, we considered functions that changed each decimal digit to another decimal digit. How about considering functions which act on decimals by shifting the decimal digits to other locations? This may be formalized as follows:

Let π be any permutation of the set $\mathbb{N}$ of natural numbers. Then π induces a well-defined function

$$\pi^* : (0,1) \to (0,1),$$

$$x = \sum_{n=1}^{\infty} a_n / 10^n \rightsquigarrow a_{\pi(n)} / 10^n.$$

Thus, $.a_1 a_2 a_3 \ldots \rightsquigarrow .a_{\pi(1)} a_{\pi(2)} a_{\pi(3)} \ldots$. For example, suppose $\pi(n) = n + (-1)^{n+1}$ for $n \in \mathbb{N}$. Then $\pi^*(.12345\overline{0}) = .2143050$, and $\pi^*(.\overline{123}) = \pi^*(.1231231231 \ldots) = .2113322113 \ldots = .\overline{211332}$. For a different function π^*, define $\pi(n)$ by $\pi(1) = 3$, $\pi(2) = 2$, $\pi(3) = 1$, and $\pi(n) = n$ for all $n \geqslant 4$. Then $\pi^*(.12345\overline{0}) = .321450$, and $\pi^*(.\overline{123}) = \pi^*(.123123123 \ldots) = .321123123 \ldots = .321\overline{123}$.

Continuity and differentiability properties of the π^* functions are considered in [3]. The reader may explore these properties in the exercises below.

Exercise 6. All π^* are continuous from above on $(0, 1)$.

[*Hint*: Show that $|\pi^*(y) - \pi^*(x)| < 10^{-k}$ for $y > x$ satisfying $|y - x| < 10^{-M}$, where $M = \max\{\pi(1), \pi(2), \ldots, \pi(k)\}$.]

Exercise 7. All π^* are continuous on $\mathfrak{N}\mathfrak{I}$.

Digit matching arguments applied to π^* may fail for $x \in \mathfrak{I}$ for reasons not unlike those that applied to p^*. One interesting subfamily to consider consists of functions π^* induced by π having $\pi(n) = n$ for all n greater than some integer $N \geqslant 0$. Denote such a function as π_N. Note, for example, that $\pi_0^*(x) = x$ on $(0, 1)$ since $\pi_0(n) = n$ for all $n \in \mathbb{N}$.

Exercise 8. π^* is differentiable at some $x \in (0, 1)$ if and only if $\pi = \pi_N$ for some $N \geqslant 0$.

[*Hint*: Assume that π^* is differentiable at $x = \sum_{n=1}^{\infty} a_n / 10^n$ and define h_n by $h_n = 10^{-n}$ for $a_n \neq 9$ and $h_n = -10^{-n}$ for $a_n = 9$. Then x and $x + h_n$ differ only in their nth digits, and $\pi^*(x)$ and $\pi^*(x + h_n)$ disagree only at the digit with index $\pi^{-1}(n)$. Since π^* is assumed differentiable at x, $\lim_{h_n \to 0} [\pi^*(x + h_n) - \pi^*(x)]/h_n$ exists and equals $\lim_{n \to \infty} 10^{n - \pi^{-1}(n)}$. Therefore, $\lim_{n \to \infty} n - \pi^{-1}(n)$ is an integer c. Show that this establishes the existence of an integer $N \geqslant 0$ such that $\pi(n) = n + c$ for all $n > N$. Conclude that the first $N + c$ natural numbers are the 1-1 images of the first N numbers, and so $c = 0$.]

Exercise 9. Every π_N^* is differentiable on $(0, 1)$ except for possible finitely many $x \in \mathfrak{I}$. Furthermore, $d\pi_N^*/dx = 1$ wherever π_N^* is differentiable.

[*Hint*:   First verify that the possible points of discontinuity for π_N are the 10^N terminal decimals $.a_1 a_2 \ldots a_j \bar{0}$ having $j \leqslant N$.]

(c) *A Noncomputable Densely Differentiable Function.*   It is well known that the set $\mathfrak{Q}$ of irrational numbers can be enumerated $r_1, r_2, \ldots, r_n, \ldots$. Define $F(x) = \sum_{r_n < x} 1/2^n$ for any real number x. The reader can verify (somewhat as in Exercise 1) that F is continuous on $\mathfrak{I}$ and nowhere continuous on $\mathfrak{Q}$. Since F is strictly increasing, it can be proved (by a theorem of Lebesgue) that F is differentiable everywhere except possibly on a set of "measure" zero (this is a form of dense differentiability). A particularly unusual feature of F is that it is noncomputable, that is, one cannot actually compute functional values $F(x)$ for $x \in \mathbb{R}$!

Remark.   This example was provided to the author in a private communication [6].

The functions described in this article can be used to sharpen readers' understanding and intuition of notions of continuity and differentiability. These functions serve as a nice antidote to the primitive notion of continuous functions being those which can be "graphed without picking up the pencil," etc.

REFERENCES

1. Lung Ock Chung, Advanced problem #6054, Amer. Math. Monthly, 82 (November 1975) 941–942.
2. Michael W. Ecker, Elementary problem #2738, Amer. Math. Monthly, 85 (November 1978) 764–765. Solution: Alberto Guzman, 87 (January 1980) 61–62.
3. Michael W. Ecker, Function pathology, Crux Mathematicorum, 5 (August-September 1979) 182–188, 190.
4. Bernard R. Gelbaum and John M. Olmsted, Counterexamples in Analysis, Holden-Day, San Francisco, 1964.
5. Brent Hailpern, Continuous nondifferentiable functions, Pi Mu Epsilon J., 6 (Fall 1976) 249–260.
6. John Mack, Australia, letter to Michael W. Ecker, November 1979.
7. Erwin Neuenschwander, Riemann's Example of a continuous "non-differentiable" function, Math. Intelligencer, 1 (1) (1978) 40–44.
8. S. L. Segal, Riemann's example of a continuous "non-differentiable" function continued, Math. Intelligencer, 1 (2) (1978) 81–82.

5. PROBABILITY & STATISTICS

Man's attempt to understand his environment and control his future requires an intelligent integration of empirical knowledge with mathematical reasoning—a blend, so to speak, of two important branches of mathematics: statistics and probability theory. Statistics is the branch of mathematics that deals with the collection, organization, and interpretation of numerical facts. Probability theory is concerned with the relative frequencies or likelihood of observed phenomena over the long run.

The origin of statistics goes back to antiquity. Governments have always gathered numerical data in order to analyze and manage the affairs of state.* We see in the Bible, for example, that censuses were taken more

---

*The term statistics is a derivation of the Latin word statisticus, meaning "of the state."

than two thousand years ago for tax-collecting and manpower-planning purposes. In its earliest form, statistics was descriptive—that is, primarily concerned with describing and summarizing data by way of numerical terms and graphical displays. Modern statistical theory, however, is inferential in nature: general conclusions are drawn from the specific—as, for example, inferring properties about the whole population from the observed content of a sample. Some of the most important developments in the foundations of modern statistical theory came about in the seventeenth century. In 1662, for instance, John Gaunt published an analysis of the recorded deaths in London from 1600–1650. Gaunt's work spawned the study of mortality rates and insurance contracts. An insurance policy, of course, is a wager between the individual insured and the company insuring that individual.

Gambling not only gave impetus to modern statistical theory, it had an even greater impact on the origin of probability theory. Indeed, some consider the Italian physician, mathematician, and avid gambler Jerome Cardan (1501–1576) to be the founder of probability theory, since his treatise "Book on Games of Chance" constituted the first theoretical study of the mathematics of chance. Although this first gambler's manual was written in 1520, it did not appear in publication until 1663. Meanwhile, in 1654, Fermat (1601–1665) and Pascal (1623–1662) began to correspond with each other on a gambling problem involving dice. Accordingly, many historians attribute the first systematic study of probabilty theory to Fermat and Pascal. In either case, it is clear that the earliest developments in probability theory were of a gambling orientation.

Statistics and probability theory have each grown and spread far beyond applications to games of chance. The most significant advances, however, combine and interlace the fundamentals of each. The sequence of hypothesis, experiment, and test of hypothesis permeates every area of prediction and decision-making. Thus, modern statistical methods, using probability theory, form the cornerstone of contemporary scientific inquiry.

The articles in this chapter illustrate some of these evolutionary developments in probability and statistics, both as individual disciplines and as a combined study.

In "Continuous Dice," Lawrence Pozsgay generalizes discrete dice in order to provide a simple, yet interesting example of a continuous distribution. Here, Pozsgay uses elementary calculus to compare and contrast probability-theoretic concepts for the discrete and continuous case. His article concludes with a game that nicely extends its discussion.

The Gambler's Ruin Problem was first investigated in the mid-seventeenth century. One solution was given by De Moivre in 1711. Other solutions are known—as, for example, Feller's method based on difference equations. In "Gambler's Ruin," Homer Hayslett, Jr., presents an intuitively appealing solution that also introduces readers to elementary notions about martingales. An interesting application of the Gambler's Ruin Problem, to the survival of the species, is also provided.

Probability theory is deductive in nature; it ascribes the likelihood of a specific sample characteristic based on the general population's properties. A good illustration of this, and the way probability-theoretic concepts interface with other branches of mathematics, is given in the article "A Graph-Theoretic Approximation of e" by Bruce Golden and Daniel Casco. Here the authors combine results from probability and graph theory in order to approximate e and to consider the expected number of endpoints of certain graphs.

Students first encounter descriptive statistics when dealing with numerical data corresponding to measurements of a single variable. If two variables are involved, the population under consideration consists of pairs of numbers—and the least-squares line is used to represent data whose scatter diagram appears linear. Since students know that the perpendicular distance from a point to a line is the shortest, they frequently wonder why distances to the least-squares line are not measured perpendicularly. David Farnsworth and Steven Suddaby answer this and more. In "The Orthogonal Least-Squares Line," they give an informative, balanced comparison of two least-squares type lines. This discussion will also help students to clarify the distinction between regression and correlation.

The classical Birthday Problem serves as an excellent vehicle for combining probability theory with statistics. Neville Spencer uses a modified version of this problem to stimulate and enrich learning in an elementary statistics course. His article "A Statistical Experiment," showing how an interesting statistical experiment can be performed in class, further illuminates the nature of statistical decision-making processes.

The St. Petersburg game, a paradox in gambling, appeared in 1725 and stymied many of the foremost probabilists of the time. Buffon, in an empirical trial, played 2048 games for an average win of 4.91 crowns per game. In "Buffon, the Computer, and the Petersburg Paradox," David Tolman and James Foster use a computer to repeat Buffon's 2048-game trial 1,000 times. The authors discovered that the limiting median, of the first n means, is a "fair" entrance fee to pay in the sense that it gives the player a 50% chance to finish an n-game series with positive cumulative winnings. This article nicely illustrates how simulation with computers can provide today's students with insight that was not available to Buffon and his contemporaries.

Continuous Dice

Lawrence J. Pozsgay, *Bridgeport, CT*

Although there are many ready-made examples to illustrate the concept of a discrete probability distribution (coins, cards, dice, etc.), most illustrations of a continuous probability distribution are either trivial (such as the uniform distribution for a freely-spinning pointer) or else extremely sophisticated (like the normal distribution). The following natural extension of discrete dice provides an example of a continuous distribution that is both simple and interesting. This illustration offers an application of elementary calculus to probability theory, and it can be useful in calculus courses for students with some knowledge of probability. In a statistics or probability course without a calculus prerequisite, one can simply present the distribution without its derivation.

Instead of a pair of dice, let Z be the continuous variable representing the sum $X + Y$ shown (Figure 1) on the two spinners after a random spin.

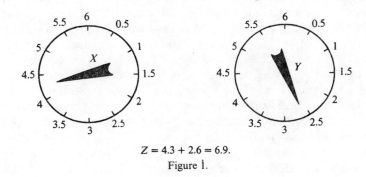

$$Z = 4.3 + 2.6 = 6.9.$$

Figure 1.

For discrete dice, one can determine the relative probabilities for the various values of Z by considering the pairs (X, Y) such that $Z = X + Y$. Figure 2a shows this for $Z = 8$. In the continuous case, we cannot consider $P(Z)$ for a particular Z (such $P(Z)$ always equals zero); instead, we must consider the probability of Z over a range of values—say $3.5 < Z \le 8.4$ (the shaded section in Figure 2b).

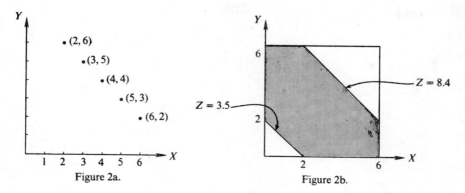

Figure 2a.   Figure 2b.

Let $A(z)$ represent the area covered by the point set $\{(X, Y) : 0 \leqslant X + Y \leqslant z\}$.

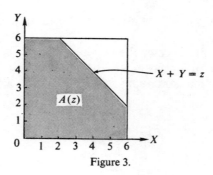

Figure 3.

It is a simple matter to verify that

$$A(z) = \begin{cases} \int_0^z \int_0^{z-x} dy\, dx & = \frac{1}{2} z^2 & \text{if } 0 \leqslant z \leqslant 6, \\ 36 - \int_{z-6}^6 \int_{z-x}^6 dy\, dx & = 36 - \frac{1}{2}(12 - z)^2 & \text{if } 6 < z \leqslant 12. \end{cases}$$

Dividing $A(z)$ by 36, we get the distribution function for continuous dice

$$F(z) = \begin{cases} 0, & z \leqslant 0, \\ \frac{1}{72} z^2, & 0 < z \leqslant 6, \\ 1 - \frac{1}{72}(12 - z)^2, & 6 < z \leqslant 12, \\ 1, & 12 < z. \end{cases}$$

The distribution functions $F(z) = P(Z \leqslant z)$ for discrete dice and for continuous dice are shown in Figure 4a and 4b, respectively.

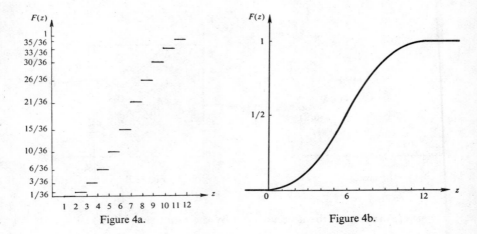

Figure 4a.   Figure 4b.

Once $F(z)$ is known, we can obtain $P(a < Z \leqslant b)$ for all a and b. Thus,

$$P(3.5 < Z \leqslant 8.4) = F(8.4) - F(3.5) = \left[1 - \tfrac{1}{72}(3.6)^2\right] - \left(\tfrac{1}{72}(3.5)^2\right) = .6499.$$

This is simply the difference of the two altitudes in Figure 4b.

Probability distributions can also be illustrated by showing their frequency curves. For the continuous case, we seek the density function $f(z)$ satisfying

$$F(z) = \int_{-\infty}^{z} f(w)\,dw$$

and obtain

$$f(z) = F'(z) = \begin{cases} 0, & z \leqslant 0, \\ \tfrac{1}{36}z, & 0 < z \leqslant 6, \\ \tfrac{1}{36}(12 - z), & 6 < z \leqslant 12, \\ 0, & 12 < z. \end{cases}$$

For the discrete case, $f(z)$ is simply $P(z)$. Figures 5a and 5b compare and contrast the density functions for discrete and continuous dice. The analogy between discrete and continuous can be further illustrated by considering the events in Figures 2a and 2b. In particular, Figures 5a and 5b show that

$$P(8) = 5/36 \quad \text{and} \quad P(3.5 < Z \leqslant 8.4) = \int_{3.5}^{8.4} f(z)\,dz = .6499$$

are the height and area of these events under their respective density functions.

The following game nicely extends the preceding discussion.

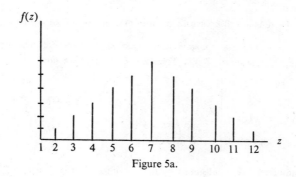

Figure 5a.

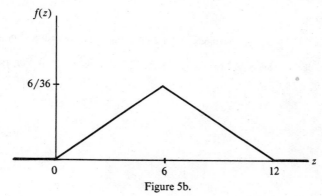

Figure 5b.

Example. If one person guesses any number $X \in [1,6]$ and another person guesses a number $Y \in [8,15]$, independently of the first player, what is $P(X + Y \leqslant z)$ for certain specified values z?

A good followup lesson, or assignment, would be to slightly generalize this example: Explore the distribution of $Z = X + Y$ for independent random variables X and Y having respective uniform distributions on $[a, b]$ and $[c, d]$, where $d - c > b - a$. The solution is straightforward. If $F(z) = P(Z \leqslant z)$, then

$$
F(z) = \begin{cases}
0, & z \leqslant a + c, \\[2mm]
\dfrac{1}{2} \cdot \dfrac{(z - a - c)^2}{(b - a)(d - c)}, & a + c < z \leqslant b + c, \\[3mm]
\dfrac{1}{2} \cdot \dfrac{(2z - 2c - a - b)(b - a)}{(b - a)(d - c)}, & b + c < z \leqslant a + d, \\[3mm]
\dfrac{(d - c)(b - a) - 1/2(z - b - d)^2}{(b - a)(d - c)}, & a + d < z \leqslant b + d, \\[3mm]
1, & b + d < z.
\end{cases}
$$

Editor's Note: There is a well-known theorem in mathematical statistics (see [1], [2]):

If X and Y are independent random variables having uniform distributions over the interval (a, b), their sum $Z = X + Y$ has the distribution function

$$F(z) = \begin{cases} 0, & z \leq 2a, \\ \dfrac{1}{2} \cdot \left(\dfrac{z - 2a}{b - a} \right)^2, & 2a < z \leq a + b, \\ 1 - \dfrac{1}{2} \cdot \left(\dfrac{2b - z}{b - a} \right)^2, & a + b < z \leq 2b, \\ 1, & 2b < z. \end{cases}$$

The case of continuous dice is subsumed by the above for $a = 0$ and $b = 6$.

REFERENCES

1. H. Cramer, Mathematical Methods of Statistics, Princeton Univ. Press, Princeton, NJ, 1946, p. 245.
2. B. V. Gnedenko, Theory of Probability, Chelsea, New York, 1967, pp. 179–181.

Gamblers' Ruin

Homer T. Hayslett, Jr., *Colby College, Waterville, ME*

The Gamblers' Ruin Problem, first investigated in the mid-seventeenth century [7], may be stated as follows:

> Gambler A *has* $a *and Gambler* B *has* $b. *The gamblers agree to bet* $1 *against each other in a sequence of bets until one of the players is ruined* (*i.e., out of money*). *What is the probability that a particular one of the two gamblers is ruined*?

Let p and $q = 1 - p$ denote the respected probabilities that A and B win any single bet. DeMoivre proved, in 1711, that the probability of A being ruined is

$$\alpha = \begin{cases} \dfrac{a}{a+b} & \text{when } p = \dfrac{1}{2}, \\[2ex] \dfrac{\left(\dfrac{q}{p}\right)^{a+b} - \left(\dfrac{q}{p}\right)^{a}}{\left(\dfrac{q}{p}\right)^{a+b} - 1} & \text{when } p \neq \dfrac{1}{2}. \end{cases} \tag{1}$$

A simple derivation of (1), using the method of difference equations, is given by Feller [4, pp. 344–345]. Our objective is to present another intuitively appealing derivation that also introduces readers to elementary notions concerning martingales in Probability Theory. A second objective is to **show how a gambler who is playing against unfavorable odds can, by varying the size of his bets, either increase his chances of winning a fixed amount or increase his playing time.** An interesting application of the Gamblers' Ruin Problem to survival of the species will also be considered.

1. Preliminary Notions

Let the random variable X_n denote A's fortune just after the nth bet and its associated payoff. It is helpful to think of a bet and its payoff being made simultaneously at unit intervals of time (each minute, perhaps). Hence, rather than referring to "A's fortune just after the nth bet and its associated payoff," we can more conveniently refer to "A's fortune at time n." This

141

has the added advantage of conforming to established terminology. Gambler A's fortunes at times $0, 1, 2, \ldots$ form a sequence of independent random variables $\{X_n\}_{n \geqslant 0} = X_0, X_1, X_2, \ldots$. (X_0 denotes A's initial fortune, namely a.) For example, if the first five outcomes of the game is the sequence $WWLWL$ (W and L respectively represent a win and a loss by A), then $X_0, X_1, X_2, X_3, X_4, X_5$ take on the values a, $a + 1$, $a + 2$, $a + 1$, $a + 2$, $a + 1$.

The sequence of bets ends when one of the gamblers is ruined. Since we do not know exactly when this will happen, the time T at which one of the gamblers is ruined is itself a random variable called the *stopping time*. Thus, A's fortune at stopping time T is

$$X_T = 0 \quad \text{or} \quad X_T = a + b, \tag{2}$$

depending on whether the game ends because A is ruined or because B is ruined.

A sequence of random variables $\{Y_n\}_{n \geqslant 0}$ is called a *martingale** if the expected value of Y_n, given the value of Y_{n-1}, is the value of Y_{n-1}. Symbolically,

$$E(Y_n \mid Y_{n-1}) = Y_{n-1} \tag{3}$$

for each natural number n. This is not the most general definition of a martingale, but it is suitable for the present discussion.

If, as in our game, Y_0 denotes the initial fortune of a gambler and Y_1 denotes his fortune after one play, the game is considered "fair" if $E(Y_1) = Y_0$. If the gambler plays the game again and Y_2 is his fortune after the second play, the game is "fair" if $E(Y_2 \mid Y_1) = Y_1$. Continuing in this manner, it becomes clear that one natural definition of fairness is the martingale condition (3). Thus, a martingale can be interpreted as a fair game.

2. Derivation of the Ruin Probabilities

The key to our derivation of (1) is the Optimal Stopping Theorem in martingale theory (see [1, p. 306]):

> *If the martingale $\{X_n\}_{n \geqslant 0}$ is bounded (i.e., there exists a constant M such that $|X_n| \leqslant M$ for all n), then*
>
> $$E(X_T) = X_0. \tag{4}$$

In order to invoke the Optimal Stopping Time, we first verify that the sequence $\{X_n\}_{n \geqslant 0}$ of A's fortunes is a martingale if and only if $p = 1/2$.

---

*Much of martingale theory has been developed by J. L. Doob. Chapter 7 of his book [2] offers the first comprehensive treatment of martingales. It appeals to one's intuitive understanding of gambling in order to illuminate the formal definitions and theorems. Those with strong mathematical backgrounds would also find [3] interesting.

This is clear since A's fortune at time n is $X_n = X_{n-1} + 1$ if he wins, and $X_n = X_{n-1} - 1$ if he loses. Thus

$$E(X_n | X_{n-1}) = p(X_{n-1} + 1) + q(X_{n-1} - 1) = X_{n-1} + (p - q),$$

and so $E(X_n | X_{n-1}) = X_{n-1}$ if and only if $p = q = 1/2$. There are two cases to consider.

(i) Suppose $p = 1/2$. Since α denoted the probability of A being ruined,

$$E(X_T) = 0 \cdot \alpha + (a + b) \cdot (1 - \alpha) = (a + b) \cdot (1 - \alpha).$$

Substituting this and $X_0 = a$ into (4), we obtain

$$\alpha = \frac{b}{a + b}.$$

(ii) Suppose $p \neq 1/2$. Although $\{X_n\}_{n \geq 0}$ is not a martingale, the sequence $\{ Y_n = (q/p)^{X_n} \}_{n \geq 0}$ is. Verification is straightforward since A's fortune at time n is $X_{n-1} + 1$ or $X_{n-1} - 1$, depending on whether he wins or loses on the nth play. Thus,

$$E(Y_n | Y_{n-1}) = p\left(\frac{q}{p}\right)^{X_{n-1}+1} + q\left(\frac{q}{p}\right)^{X_{n-1}-1} = \left(\frac{q}{p}\right)^{X_{n-1}} = Y_{n-1}.$$

Since X_T assumes only the values 0 (with probability α) and $a + b$ (with probability $1 - \alpha$), $Y_T = (q/p)^{X_T}$ can only assume the values $(q/p)^0 = 1$ or $(q/p)^{a+b}$. Therefore

$$E(Y_T) = \alpha \cdot 1 + (1 - \alpha)\left(\frac{q}{p}\right)^{a+b},$$

and (4) becomes

$$\alpha + (1 - \alpha)\left(\frac{q}{p}\right)^{a+b} = \left(\frac{q}{p}\right)^a.$$

Solving this equation for α, we get

$$\alpha = \frac{\left(\frac{q}{p}\right)^{a+b} - \left(\frac{q}{p}\right)^a}{\left(\frac{q}{p}\right)^{a+b} - 1}.$$

In addition to knowing the probability that A will be ruined, another question of interest is: *How long will the game last?* Since the stopping time T (i.e., the time when the game ends) is a random variable, it is not possible to know in advance exactly how long a specific game will last. But it is possible to find an expression for $E(T)$, the expected duration of the game. This, of course, depends on the fortunes of A and B, as well as the probability p that A wins a single bet. Let $D(a, b; p)$ denote $E(T)$ for the game where A and B begin with \$$a$ and \$$b$, and p is the probability of A winning a single play. The reader can readily prove (or refer to [4, pp. 388–389]) that

$$
D(a,b;\,p) = \begin{cases} ab & \text{when } p = \dfrac{1}{2}, \\[2em] \dfrac{a}{q-p} - \dfrac{a+b}{q-p} \left[\dfrac{\left(\dfrac{q}{p}\right)^{a} - 1}{\left(\dfrac{q}{p}\right)^{a+b} - 1} \right] & \text{when } p \neq \dfrac{1}{2}. \end{cases}
\tag{5}
$$

3. Examples and Discussion

For convenience, let $\alpha(a,b;\,p)$ denote the probability of A being ruined in the game where A and B begin with $\$a$ and $\$b$, and p is the probability of A winning a single bet.

Example 1. Suppose $p = \frac{1}{2}$. Then $\alpha = (a,b;\frac{1}{2}) = \frac{1}{2}$ and $\alpha(\frac{3}{2}b,b;\frac{1}{2}) = \frac{2}{5}$, no matter how much a and b are. In general, when $a = xb$ ($x > 0$), formula (1) yields the fixed value $\alpha(xb,b;\frac{1}{2}) = 1/(1+x)$, no matter how large a and b are. The expected duration of the game, however, does depend on a and b, as well as on x. For instance, (5) reveals that $D(xb,b;\frac{1}{2}) = b^2 x$. Thus for $x = \frac{1}{2}$, we have

$$
D(1,2;\tfrac{1}{2}) = 2 \quad \text{and} \quad D(2,4;\tfrac{1}{2}) = 8, \quad \text{whereas } D(20,40,\tfrac{1}{2}) = 800.
$$

Example 2. One version of the Gamblers' Ruin Problem, published by Christian Huygens in 1657, is stated as follows:

> "A and B take each twelve counters and play with three dice on the condition that if eleven is thrown, A gives a counter to B, and if fourteen is thrown, B gives a counter to A; and he wins the game who first obtains all the counters. Show that A's chance is to B's as $244,140,625$ is to $282,429,536,481$."

It may be an instructive exercise for readers to calculate the probabilities $\alpha(12,12;\,p_r(11))$ and $\beta(12,12;\,p_r(14))$, and odds, of A and B being ruined.

Example 3. Suppose $p = .45$. Making use of a calculator, we find from (1) that

$$
\begin{aligned}
\alpha(1,1;.45) &= .55 \\
\alpha(2,2;.45) &= .60 \\
\alpha(4,4;.45) &= .69 \\
\alpha(20,20;.45) &= .98 \\
\alpha(100,100;.45) &= .99999.
\end{aligned}
$$

Thus, it is almost certain that A will be ruined when $a = b$ is large.

The probabilities that A will be ruined increase even more rapidly when A has fewer dollars than B. (This would certainly be the case, for example, when A is an individual gambler and B is a casino.) If A has only half as many dollars as B, then we calculate that

$$\alpha(1,2;.45) \quad = .73$$
$$\alpha(2,4;.45) \quad = .79$$
$$\alpha(4,8;.45) \quad = .88$$
$$\alpha(20,40;.45) = .99968.$$

The corresponding durations, obtained from (5), are

$$D(1,2;.45) = 1.93$$
$$D(2,4;.45) = 7.30$$
$$D(4,8;.45) \quad = 25.39$$
$$D(20,40;.45) = 199.8.$$

Some intriguing opportunities for being able to effect ruin probabilities and the duration of the game arise from inspection of the above figures.

In Example 3, for instance, suppose that A has $20 and B has $40, and that A is desperate to have $60. If each gambler bets $1 at a time, then (as we have previously calculated) the probability that A will be ruined (rather than obtain $60) is .99968. Suppose however that A can convince B that they should bet $5 each time instead of $1. Then the situation is the same as if their fortunes were 4 and 8, instead of 20 and 40, and the probability that A will be ruined drops to .88. If the gamblers wager $20 on each bet, it is as if their fortunes were 1 and 2, and the probability of A's ruin drops still further to .73. Thus, we see that A has increased his probability of attaining $60 from .00032 to .27. In general: *The best win strategy, when gambling against unfavorable odds and against an opponent with more resources, is to bet as much as possible on each individual bet.*

On the other hand, A might not care much whether he is ruined or not, but merely might want to play as long as possible. (That is, A would be thinking of gambling as pleasurable and would look upon any money lost simply as payment for that pleasure.) *Then A can increase the expected duration of the game by decreasing the size of the bet.* For example, if $p = .45$ and A has $1 and B has $2, the expected number of bets until one of the gamblers is ruined (Gambler A 73% of the time) is 1.93. If, instead of betting the entire $1, A bets only a quarter each time, A can increase the expected duration of the game to 25.39 bets (the fortunes now being 4 and 8), but A has also raised the probability of his ruin from .73 to .88. The expected duration of the game can be increased still further by again decreasing the bet: If A bets only a nickel each time, the expected duration of the game becomes 199.8 (the fortunes being 20 and 40), and the probability of A's ruin is raised to .99968. Therefore if A's desire is to play as long as possible, and he is indifferent to being ruined, the best strategy is to reduce his bet on each play to the lowest possible level.

4. Gamblers' Ruin in Biology

The preceding strategy is considered in an interesting biological context by Slobodkin and Rappaport [6]. They view the evolutionary response of a species to environmental change as a game between the species and nature. One way of looking at individuals' and species' behavioral and genetic responses to environmental change is to consider them as bets which may or may not be successful. The species, if it wins the bet (i.e., successfully adapts to some change), cannot take the payoff (i.e., evolutionary success) and leave the game: the payoff only allows the species to continue playing the game. Thus, the reward for winning constitutes being able to continue playing the game. (This kind of game is called an existential game.) The individual or species would like to continue playing the game as long as possible. In order to do so, as we have just seen in the above paragraph, a player should bet as little as possible. We can think of the bets made by the species as changes in its genetic make-up. Therefore, in order to continue playing the game as long as possible (remaining in existence), the individual or species should change as little as possible in response to any change in its environment.

REFERENCES

1. Kai Lai Chung, Elementary Probability Theory with Stochastic Processes, Springer-Verlag, New York, 1974.
2. J. L. Doob, Stochastic Processes, Wiley, New York, 1953.
3. ———, What is a martingale?, Amer. Math. Monthly, 78 (1971) 451–462.
4. William Feller, An Introduction to Probability Theory and its Applications, Vol. 1, 3rd ed., Wiley, New York, 1968.
5. L. E. Maistrov (Samuel Kotz, Editor and Translator), Probability Theory: A Historical Sketch, Academic Press, New York, 1974.
6. Lawrence B. Slobodkin and Anatol Rappaport, An optimal strategy of evolution, Quart. Rev. Biology, 49 (1974) 179–200.
7. Lajos Takacs, On the classical ruin problems, J. Amer. Statist. Assoc., 64 (1969) 889–906.

A Graph-Theoretic Approximation of *e*

Bruce Golden and Daniel Casco, *University of Maryland, College Park, MD*

One straightforward method for estimating *e* is to make use of the infinite series

$$e = 1 + \frac{1}{1!} + \frac{1}{2!} + \frac{1}{3!} + \frac{1}{4!} + \cdots .$$

A more interesting and somewhat unusual approach for approximating *e* is possible by combining basic results from probability and graph theory.

A *graph* is a set of points (called *nodes*) and a set of lines (called *arcs*) which connects nodes. A graph is *complete* if every pair of its nodes is connected by an arc. The degree of a node is the number of arcs which touch it. Accordingly, a node of degree one is called an *endpoint*. A *spanning tree on n nodes* is a graph with $n - 1$ arcs and no cycles (i.e., no collection of arcs which form a closed polygon with nodes as vertices). We illustrate these concepts in Figures 1, 2, and 3. For instance, in Figure 3, nodes 2, 4, and 5 are endpoints, whereas the degree of node 1 is three.

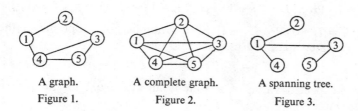

A graph.	A complete graph.	A spanning tree.
Figure 1.	Figure 2.	Figure 3.

Cayley's well-known theorem*

> *The number of spanning trees of the complete graph K_n on n nodes is $T(n) = n^{n-2}$*

---

*This is proved in a number of texts. See, for example, J. Riordan, *An Introduction to Combinatorial Analysis*, Wiley, 1958, and J. Moon, "Various Proofs of Cayley's Formula for Counting Trees," in *A Seminar on Graph Theory*, Holt, 1967.

leads to the following observation about a spanning tree randomly selected from the set of all possible spanning trees:

For large n, the probability that a given node of a spanning tree of K_n is an endpoint is approximately e^{-1}.

The total number of spanning trees of K_n is $T(n) = n^{n-2}$, and the total number of K_{n-1} is $T(n-1) = (n-1)^{n-3}$. Choose a particular node m as an endpoint. Since K_n is complete, the removal of node m and the arcs which touch it leaves a complete graph K_{n-1}. There are $T(n-1)$ spanning trees of K_{n-1}. Each of these spanning trees can be connected to node m by connecting m to one of the remaining $(n-1)$ nodes. Thus, the number of spanning trees of K_n with m as endpoint is $(n-1)T(n-1)$. The probability that node m is an endpoint of a spanning tree of K_n is

$$ p_n = \frac{(n-1)T(n-1)}{T(n)} = \frac{(n-1)(n-1)^{n-3}}{n^{n-2}} = \left(1 - \frac{1}{n}\right)^n \cdot \left(\frac{n}{n-1}\right)^2. $$

Since $\lim_{n \to \infty}(1 - 1/n)^n = e^{-1}$, we see that p_n approaches e^{-1} as n gets large.

The expected number of endpoints in a randomly generated spanning tree of K_n is np_n. For large n, this expected number is approximately n/e. To see this, select a spanning tree at random from the $T(n)$ possible spanning trees, and define

$$ I_j = \begin{cases} 1 & \text{if node } j \text{ is an endpoint in the spanning tree,} \\ 0 & \text{otherwise.} \end{cases} $$

Since the probability that a given node is an endpoint is p_n, we have the expected value of I_j

$$ E(I_j) = 0 \cdot \text{Prob}\{I_j = 0\} + 1 \cdot \text{Prob}\{I_j = 1\} = 0 \cdot (1 - p_n) + 1 \cdot (p_n) = p_n. $$

Letting $S = I_1 + I_2 + \cdots + I_n$ be the number of endpoints in the randomly generated spanning tree, we find that

$$ E(S) = E(I_1) + \cdots + E(I_n) = np_n $$

since the expected value of a sum is always equal to the sum of the expected values.

The Orthogonal Least Squares Line

David Farnsworth, *Eisenhower College, Seneca Falls, NY,* and Steven Suddaby, *Administrative Office of the United States Courts, Washington, DC*

Students in elementary statistics courses often ask, "Why aren't the distances to the least squares line measured perpendicular to the line?" They point out that the perpendicular distance is the shortest and is the one they used in their elementary mathematics courses.

Teachers find the same problem when comparing grades on two exams for a group of students. The question might be posed as, "Which should be x and which y?"

These questions can be answered with a discussion of ordinary least squares lines and orthogonal least squares lines.

Ordinary Least Squares Lines

The ordinary least squares or regression line is used to model data which in its scatter diagram appears to be linear. The usual formulae for a and b in terms of the n data points $\{(x_i, y_i): 1 \leqslant i \leqslant n\}$ for the line $y = a + bx$ carry with them some assumptions. The major one is that each x-measurement is absolutely precise but that the y-measurements each have an associated error or standard deviation δ_y. An estimator of δ_y is the standard deviation of repeated y-measurements at a fixed x. It is different from the standard deviation σ_y, the measure of variability among all members of the population of possible observations y.

Theoretically, there are two lines which coincide if and only if the data lie exactly on a line. The second line is $x = c + dy$ and, except in the exactly linear case, is not obtainable from $y = a + bx$ by solving for x. This second line is used correctly if the assumption that there is no error in each y-measurement (so $\delta_y = 0$) is satisfied, and that there is a standard deviation δ_x associated with each x-measurement. Both lines contain the point $(\bar{x}, \bar{y})$. It should be pointed out that although in theory two lines exist and it is fruitful to discuss them here, only one of them—if any—is appropriate in ordinary practical situations. Normally the y-variable is chosen as the one to be predicted, and it is assumed that $\delta_x = 0$. Examples for which $\delta_x = 0$ abound in elementary statistics books.

Consider the artificial data $(0,1)$, $(1,2)$, $(1,3)$, $(2,2)$, $(2,3)$ and $(3,4)$. The usual least squares line, where δ_x is assumed to be zero and the sum of the squared distances measured parallel to the y-axis is minimized, is

$$y = 1.27 + 0.82x \tag{1}$$

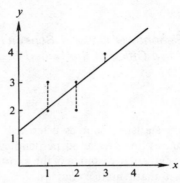

The usual least squares (i.e., regression) line with distances from the data to the line indicated with dotted lines.

Figure 1.

(Figure 1). The other least squares line, where δ_y is assumed to be zero and the sum of the squared distances measured parallel to the x-axis is minimized, is

$$x = -0.55 + 0.82y. \tag{2}$$

Notice that solving (2) for y gives $y = 0.67 + 1.22x$ which is different from (1) (Figure 2). Neither of these lines has slope one even though the data is symmetric about the line $y = 1 + x$.

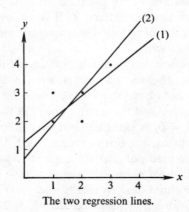

The two regression lines.

Figure 2.

The additional assumptions, such as the one that the repeated y-measurements at any fixed x would be normally distributed for the usual line with $\delta_x = 0$, lead to a t or F statistic which tests the true linearity of the phenomenon. The same t or F is obtained for both lines. Since the angle between the two lines is zero only for a perfect fit, the size of this angle is also a measure of goodness-of-fit. The r statistic for the above data is 0.818 and r^2 is 0.669. The r^2 statistic, which is always between 0 and $+1$, is another measure with $r^2 = 0$ indicating no linear relationship and $r^2 = 1$ a perfect linear relationship. This r^2 also has a reduction in sum of squares interpretation: $1 - r^2$ is the ratio of the sum of squared distances of the data points from the regression line (called residual sum of squares) to the sum of the squared distances of the data points from the mean. It measures goodness-of-fit in the y-coordinate direction which is the direction of prediction.

With the above assumption, confidence intervals and hypothesis tests can be constructed and are in common use. If the usual line is used when indeed $\delta_x \neq 0$ along with $\delta_y \neq 0$, then the slope calculated from the data will be biased towards zero (Bloch 1978). This means that any interpolations or predictions tend to be systematically in error. For lines of positive slope, say, estimates of y from x would be likely to be too low for $x > \bar{x}$ and too high for $x < \bar{x}$. This use of a regression line is like pounding nails with a wrench: you may get some result, but it could be better if you used the proper tool.

Orthogonal Least Squares Line

If there is a variance associated with both the x- and the y-measurements, that is $\delta_x \neq 0$ and $\delta_y \neq 0$, then an ordinary least squares line is not appropriate. If the assumption that $\delta_x = \delta_y$ is tenable, then the orthogonal least squares line is the one to use (first pointed out by Gini 1921).

In studying, for example, the relationship between the IQ's of monozygotic twins, it is not unreasonable to assume that $\delta_x = \delta_y$. For each pair we will designate the first-born's IQ to be the x variable and the second born's the y. It is assumed that the same intelligence test is given to all twins. The assumption that $\delta_x = \delta_y$ means that if each twin of a pair were given repeated independent tests, then the standard deviations of the x's and of the y's would be roughly equal. If the same instruments are being used for the x- and y-measurements, $\delta_x = \delta_y$ is frequently satisfied.

For this line the sum of the n squared distances measured perpendicular to the line is minimized. The slope and the point $(\bar{x}, \bar{y})$ on the line can be found by minimizing

$$D = \frac{1}{b_1^2 + 1} \sum_{i=1}^{n} (b_1 x_i + b_0 - y_i)^2$$

with respect to b_0, the y-intercept, and b_1, the slope. A good exercise for the more advanced student is to show that D is indeed the sum of the squares of the perpendicular distances from the points to the line and then to minimize it using ordinary differential calculus techniques. Defining the relationships

$$\sum_{i=1}^{n} x_i = n\bar{x}, \qquad \frac{1}{n}\left[\sum_{i=1}^{n} x_i^2 - n\bar{x}^2\right] = s_x^2,$$

and

$$\frac{1}{n}\left[\sum_{i=1}^{n} x_i y_i - n\bar{x}\bar{y}\right] = s_{xy} = rs_x s_y,$$

the slope of the line can be written as

$$b_1 = \frac{s_y^2 - s_x^2 \pm \sqrt{\left(s_y^2 - s_x^2\right)^2 + 4r^2 s_x^2 s_y^2}}{2rs_x s_y}.$$

Note that for the artificial data above (Figure 3) the slope is one. In the familiar slope-intercept form, the equation is written as $y = b_1 x + (\bar{y} - b_1\bar{x})$.

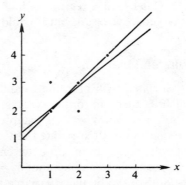

The usual regression line from Figure 1 and the orthogonal least squares line.

Figure 3.

The use of the orthogonal least sqaures line does not require that δ_x be known, but only that $\delta_x = \delta_y$. Remember that δ_x and s_x are quite different.

Additional assumptions such as the one that the data is bivariate normal lead to tests of significance and confidence intervals (Gini 1921, Villars and Anderson 1943, Lindley 1947, Creasy 1956, Moran 1956, Jolicoeur 1968, and Anderson 1975).

The orthogonal least squares line is natural for several reasons. There exist errors in both the x- and the y-measurements in most real data, for example, when x and y are continuous. The presence of a nonzero δ_x may be revealed by the lack of precision of the measuring instrument (for example, a meter stick with centimeters as its smallest units can only give measurements precise to the nearest 0.5 cm) or other uncertainties.

Measuring the distances perpendicular to the line is the normal (no pun intended) way for students in elementary mathematics courses to proceed, and this is the usual concept for physics students where the line balances a system of coplanar points. Physically, the line passes through the centroid $(\bar{x}, \bar{y})$, and is the line about which the moment of momentum or angular momentum is a minimum (Pearson 1901).

There is a strong visual argument in that this line is the axis of symmetry of the data. This is displayed with the artificial data (Figure 3). More technically, this line is the axis of the so-called correlation ellipse which we shall not discuss here (Pearson 1901 and Cramér 1946).

Comparisons Between Regression Lines and Orthogonal Least Squares Lines

For the artificial data, the regression line (1) has slope closer to zero than does the slope of the orthogonal least squares line. This is true in general and displays a clear difference between them for estimating. Hence assumptions concerning δ_x are important.

The orthogonal least squares line lies within the smaller angle between the two regression lines (Pearson 1901 and Coleman 1932).

The sum of squared distances from the orthogonal least squares line is less than or equal to the residual sum of squares for either regression line (Pearson 1901 and Coleman 1932). These distances are all measured in different directions. If we use a reduction in sum of squares criterion, this line is always better than either regression line.

Although the first publications concerning the orthogonal least squares line appeared about one hundred years ago (Adcock 1878 and Kummell 1879), it has seen some mathematical statistics uses but few applications to real data. There are a number of reasons for this. Perhaps the strongest reason is the fact that the orthogonal least squares line is not well known. Tests of significance and other useful results are in the literature but not in textbooks. Another factor is that the ordinary regression line has so many attractive properties (unbiasedness, ease of calculation, readily available confidence and prediction intervals) that it is commonly used. When δ_x is zero or very small compared to δ_y, as is frequently the case, the ordinary regression line is correct.

The problem is not just a case of a lack of publicity, however; the orthogonal least squares line has characteristics which in many circumstances are deficiencies. It is very difficult to calculate for more than one independent variable. The scope of its applicability can be rather narrow because of the $\delta_x = \delta_y$ assumption.

The predictive capability of the orthogonal least squares line is somewhat in question. Few authors have discussed this problem (exceptions include Lindley 1947 and Madansky 1959). Sometimes estimates of the slope and the y-intercept are the goal. That is, the true functional relationship between x and y is the important information (Madansky 1959). An ordinary regression line is easily interpretable as yielding an estimated or forecasted y-value for a given precise x.

Use of the Orthogonal Least Squares Line in Class

This line is an alternate choice of model for data. The model is easily accessible to students either at the level of acceptance of formulae or of deriving them. The alternative makes clearer the somewhat different concepts used in regression lines. Proofs and derivations are not presented here because they are available in the literature (Reed 1921, Coleman 1932, Lindley 1947, and Madansky 1959) or can be easily done afresh. The bibliography is intended to be comprehensive but not complete.

The distinction between regression and correlation is often clouded in classroom presentations. In regression only y, say, is random (i.e., $\delta_x = 0$); there is a fitted line $y = a + bx$, and r^2 has the interpretation as the ratio of sum of squares. For correlation analysis, both x and y are random; there is no line, and there is a population correlation coefficient ρ which is estimated by r. The orthogonal least squares line blends the two concepts of regression and correlation since it is a natural line which can be used to examine data when correlation analysis is the appropriate tool.

REFERENCES

R. J. Adcock, A problem in least squares, The Analyst (Des Moines), 5 (1878) 53–54 and 149–150.

Peter O. Anderson, Large sample and jackknife procedures for small sample orthogonal least squares inference, Comm. Statist., 4 (1975) 193–202.

Farrell E. Bloch, Measurement error and statistical significance of an independent variable, Amer. Statist., 32 (1978) 26–27.

J. B. Coleman, A coefficient of linear correlation based on the method of least squares and the line of best fit, Ann. Math. Statist., 3 (1932) 79–85.

H. Cramer, Mathematical Methods of Statistics, Princeton Univ. Press, 1946, 275–285, 309–311.

Monica A. Creasy, Confidence limits for the gradient in the linear functional relationship, J. Roy. Statist. Soc. Ser. B, 18 (1956) 65–69.

Corrado Gini, Sull' interpolazione di una retta quando i valori della variabile indipendente sono affetti da errori accidentali, Metron, 1 (1921) 63–82.

Pierre Jolicoeur, Interval estimation of the slope of the major axis of a bivariate normal distribution in the case of a small sample, Biometrics, 24 (1968) 679–682.

Charles H. Kummell, Reduction of observation equations which contain more than one observed quantity, The Analyst (Des Moines), 6 (1879) 97–105.

D. V. Lindley, Regression lines and the linear functional relationship, J. Roy. Statist. Soc., Supp. 9 (1947) 218–244.

Albert Madansky, The fitting of straight lines when both variables are subject to error, J. Amer. Statist. Assoc., 54 (1959) 173–205.

P. A. P. Moran, A test of significance for an unidentifiable relation, J. Roy. Statist. Soc. Ser. B, 18 (1956) 61–64.

Karl Pearson, On lines and planes of closest fit to systems of points in space, Philos. Mag. Ser. 6, 2 (1901) 559–572.

Lowell J. Reed, Fitting straight lines, Metron, 1 (1921) 54–61.

D. S. Villars and T. W. Anderson, Some significance tests for normal bivariate distributions, Ann. Math. Statist., 14 (1943) 141–148.

A Statistical Experiment

Neville Spencer, *California State College, San Bernadino, CA*[†]

One of the most interesting paradoxes in elementary mathematics is the classical Birthday Problem, which can be stated as follows:

> *In a given group of M people, what is the probability that two or more have the same birthdate?*

(Variations and discussion of this problem appear in [1], 142–143; [2], 380–382; and [4], 8 and 48.) The problem, when solved for different values of *M*, serves as an excellent example of how to use complementary and conditional probabilities in computation. This characteristic, together with its entertaining paradoxical appeal, makes the problem very useful in teaching a unit in elementary probability. However, in a modified form, the problem can be used in an equally appealing way in an elementary statistics course.

The form of the problem, as stated above, does not lend itself easily to experiment. However, R. S. Kleber [3] has given a modification which preserves the paradoxical nature of the original problem and makes it easier to use in the classroom. This version may be stated as follows:

> *If M selections are made randomly and independently from the set of whole numbers between 1 and N, what is the probability that at least two of the selections are the same?*

Table 1 indicates the maximum value for *N* to guarantee a probability *P* of success for *M* selections. For example, if there are *M* = 25 students in the class each making one selection, and you wish a 75% success probability, then choose *N* = 224. For the Birthday Problem analog, one would need to find *M* and *P* for *N* = 365 (which is not in Table 1). Clearly, for a probability of .5, the value *N* = 365 would have to occur between *M* = 20 and *M* = 25, if the middle column of Table 1 is examined. That is, we would have an even chance of at least one birthday duplication if there are *M* persons, where 20 < *M* < 25. With this version of the original problem,

---

[†]The author is deceased.

we have the groundwork for an interesting statistical experiment which can be performed in a reasonably short time.

Table 1.

M \ P	0.10	0.25	0.50	0.75	0.90
5	96	36	16	8	6
10	430	159	68	35	22
15	1,001	369	156	80	50
20	1,809	666	280	143	89
25	2,855	1,051	441	224	138
30	4,138	1,521	637	323	199
35	5,658	2,079	869	440	270
40	7,416	2,724	1,138	575	352
45	9,411	3,456	1,443	729	445
50	11,643	4,274	1,783	900	548

The experiment goes something like this. Prior to meeting class, the instructor chooses a value for N which yields a reasonably high probability of success for the number M of students in the class. He then poses the problem to the class, using the number N selected, and the class members guess and discuss their intuitive responses until they can agree on a probability of success—or at least an upper bound for one. This is the hypothesis to be tested. To begin the experiment, the class decides on a sample size. (This can be varied to illustrate the effect of sample size on the results and to demonstrate both large and small sample procedures.) Then the class decides on an efficient method (which is both random and independent) for selecting the numbers in each sample. (This is a good place to discuss sampling errors which occur from having the experimenters think up numbers, or from having the numbers given orally.) Now the sampling can be done, the results tallied, and the statistical testing procedure applied. The results should provide excellent motivation for students to want to know the value of the actual probability. To illustrate how this experiment succeeded in the classroom, the results of a mock experiment will be described. It was run in an imaginary class of size $M = 15$, selecting from positive integers up to $N = 80$.

Upon being given the problem, the class made estimates and finally agreed that the probability P of successfully matching at least two numbers was certainly no greater than 25%. So the null hypothesis was $H_0 : P < .25$, and the alternate hypothesis was $H_A : P \geqslant .25$. To test this hypothesis, a significance level of $\alpha = .01$ was chosen. It was also decided to choose samples of size 25, 50, and 100 to examine the effect of sample size on the confidence interval. In order to save class time, and to demonstrate the use of the computer, a table-top computer was used to generate the samples. The computer was programmed to select positive integers randomly for each trial until it had selected 15 numbers without a match or until a match occurred, at which time it proceeded to the next trial.

We considered each set of 15 random integers to be a trial in which "success" constituted obtaining *at least one* match, while "failure" meant that no match occurred. The trials with no match are marked with an asterisk. It can be seen, in Table 2, that there were 6 failures and 19 successes in the sample consisting of 25 trials. We subsequently repeated the experiment for samples of 50 and 100 trials, respectively.

Table 2.

Trial Number	Trial Results													
1	22	39	39											
2	76	61	43	70	28	60	2	61						
*3	36	3	55	37	31	48	46	77	79	70	58	65	34	57	5
4	42	29	51	57	5	59	26	74	19	30	54	28	78	78
5	30	6	78	80	37	37								
6	58	74	21	35	63	80	26	40	1	66	20	19	40	
*7	25	47	75	29	18	4	48	30	36	10	65	61	17	72	49
8	8	3	61	38	24	20	28	62	45	3				
9	40	17	65	39	59	24	2	79	50	42	65			
10	51	3	62	59	43	54	27	24	38	1	11	41	18	11
11	50	43	60	1	45	33	28	3	21	55	9	79	57	75	28
12	48	56	42	78	66	22	28	59	42					
*13	67	2	14	16	20	42	71	63	68	33	15	40	11	53	47
14	74	22	5	72	49	12	73	21	63	80	39	71	60	4	74
*15	24	47	59	48	37	39	79	68	38	44	10	4	9	75	46
16	50	25	46	57	76	76								
17	65	52	76	71	71									
18	71	58	1	26	62	73	8	27	32	55	13	31	80	27
19	34	77	63	33	32	63								
20	53	37	48	39	50	22	23	58	62	54	12	32	32	
21	74	1	19	30	35	65	51	76	53	33	1			
22	76	63	80	31	13	6	58	80						
*23	38	16	18	55	13	47	76	70	9	19	34	80	29	36	37
*24	26	61	36	4	79	69	73	25	33	39	44	31	59	24	8
25	47	55	26	15	38	71	69	75	36	35	1	73	4	55

The number of successes in each sample was tallied and the results are shown in Table 3 below.

Table 3.

	Sample 1	Sample 2	Sample 3
Sample Size (n_i)	25	50	100
Number of Successes	19	36	70
Proportion of Successes	0.76	0.72	0.70

The class used the following six-step procedure for hypothesis testing.

1. Determine the null hypothesis H_0 to be tested and the alternate hypothesis H_A;

2. Set a significance level α;
3. Determine the kind of distribution to be used;
4. Determine whether the test is one-tailed or two-tailed and find the critical value from the appropriate table;
5. Standardize the data score(s) for comparison with the critical value(s);
6. Compare the standardized data score(s) with the critical values and make the decision.

Having already determined H_0 and H_A, and a significance level $\alpha = .01$, the class proceeded to step 3. After considerable discussion, they finally decided that each sample of size n_i constituted a binomial experiment of n_i independent trials, with the probability P of success taken to be less than or equal to .25 by hypothesis. Next came the question of whether to use the binomial distribution directly for the testing procedure or to use the normal distribution as an approximation. As a guide to making this decision, the class computed the values of $P \cdot n_i$ and $(1 - P) \cdot n_i$ for each sample to determine whether or not they were greater than 5, in which case the normal distribution would be used as an approximation. Since the actual value of P was not known, the class used $P = .25$ as an estimate. The only questionable situation arose for $n_i = 25$, where $Pn_i = 6.25$. Since the actual value of P might be less than .25, the sample of size 25 posed a borderline case. It was decided to treat this sample separately using the binomial distribution directly at a later time. The other two samples were well within the guidelines; so for these samples, the tests would be performed by using the normal approximation to the binomial distribution.

For step 4, it was determined that the test was right-tailed and, for $\alpha = .01$, a table of areas under the normal curve (with two-place accuracy) yielded a critical value of $Z_{.01} = 2.33$.

The next step was to standardize the sample proportions (listed in Table 3). Using the formulas $Z = (p - P)/\sigma_p$ with $\sigma_p = \sqrt{P(1 - P)/n_i}$ (where p is the sample proportion, $P = .25$ by hypothesis and n_i is the size of the ith sample) the results were obtained as indicated in Table 4.

Table 4.

	Sample 2	Sample 3
n_i	50	100
σ_p	0.06	0.04
Z	7.67	10.39

Upon comparing these scores with the critical value of $Z_{.01} = 2.33$ as the final step in the process, the conclusion was to reject the null hypothesis. It was pointed out that, in light of the great difference between either of these scores and the critical value, there was little risk of error in accepting the alternate hypothesis that $P \geq .25$—probably by a significant amount.

To obtain an interval estimate for the actual value of P, we obtained a confidence interval using the sample of size 100 for greatest accuracy, and $\alpha = .05$ for a shorter interval. With the sample proportion $p = .70$ as an estimate for P (from Table 3), we computed

$$\sigma_p = \sqrt{\frac{(.70)(.30)}{100}} = .05.$$

With a critical value of $Z = 1.96$, we then computed the 95% confidence limits, obtaining $.70 \pm .10$ or a 95% confidence interval $0.60 < P < 0.80$. (The theoretical value of P is 0.75 as can be seen from Table 1.)

The value of performing actual experiments in an introductory statistics course is well known to experienced teachers. This approach helps to clarify vague concepts for students and makes them more meaningful. Often, peripheral questions arise which lead to other interesting problems. One such question from this experiment was motivated by noticing the different lengths of the computer trials in Table 2 and wondering what the average row length might be. A natural followup might involve the variability of the row lengths, etc.

This concludes the description of our experiment. The special sample size 25, which was to be tested using a binomial distribution, can serve as an exercise. This exercise might even be more interesting if a smaller sample were tested in this way and the results compared with those calculated for the larger samples. However, after students have performed this experiment in class, they should be curious about the actual value of P. If preparation has been made in advance, it is an excellent time to calculate it using the probabilistic method. Therefore, the experiment can serve not only to illuminate the nature of the statistical decision-making process, but also as motivation for a topic in probability.

REFERENCES

1. David Bergamini, Mathematics, Time, New York, 1963, pp. 142–143.
2. Harold R.. Jacobs, Mathematics: A Human Endeavor, W. H. Freeman, San Francisco, 1970, pp. 380–382.
3. R. S. Kleber, A Classroom Illustration of a Nonintuitive Probability, Math. Teacher, 62 (May 1969) 361–362.
4. Frederick Mosteller, Fifty Challenging Problems in Probability with Solutions, Addison-Wesley, Reading, MA, 1965, pp. 8, 48.

Buffon, the Computer and the Petersburg Paradox

David C. Tolman and James H. Foster, *Weber State College, Ogden, UT*

The Petersburg Paradox first appeared in 1725 in the Commentarii of the Petersburg Academy. It may be stated as follows:

> *Suppose Peter and Paul agree to play a game based on the flip of a coin, and Paul agrees to pay Peter one crown if the first toss comes up heads, two crowns if a head first appears on the second toss and, in general, 2^{k-1} crowns if heads first appears on the kth toss. How much should Peter pay Paul for the privilege of playing the game?*

Peter's mathematical expectation, given by

$$E = 1/2 + (1/2)^2(2) + \cdots + (1/2)^k(2^{k-1}) + \cdots$$

is clearly infinite. Thus, although seemingly contrary to common sense, if Paul's resources are infinite and an unlimited time is available for playing the game, no finite fee would produce a fair game. However, since the probability of winning a large sum is small and since one often thinks more in terms of winning or losing than in terms of the margin of win or loss, we think that a gambler (as well as a mathematician) might be interested in the following reformulation of the problem:

> *Suppose Peter agrees in advance to play Paul n Petersburg games. What fee would give Peter a 50% chance to finish the series with positive cumulative winnings?*

Our consideration of this problem will give an interesting insight into the nature of the Petersburg Paradox.

Buffon and the Computer ($n = 2048$). Buffon ([2], 185), in an empirical trial, played 2048 games producing an average winning of 4.91 crowns per game. Buffon's trial of 2048 games was later repeated by a student of De Morgan's ([2], 185), producing 53,238 crowns for an average winning of 26.0 crowns per game.

Using a computer and a random number generator,[1] we simulated the Petersburg game and repeated Buffon's 2048-game trial 1000 times. Each time we computed the average winning per game for the 2048-game series, giving us 1000 averages. The highest average was 297 crowns per game, the lowest was 4.09, and the mean was 9.81. The median, which is the statistic of interest in solving our problem, was 6.84 crowns per game. A histogram of the 1000 averages obtained is given in Figure 1.

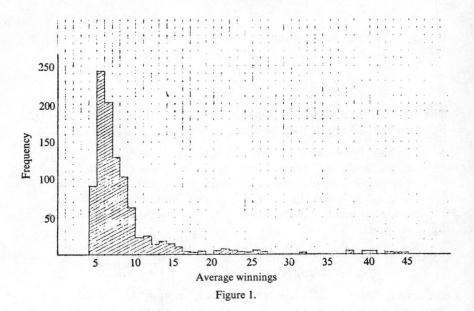

Figure 1.

Now suppose that we had not stopped at 1000 repetitions of Buffon's 2048-game series but had continued to calculate the addended mean and median as each new 2048-game average was added to the distribution. Limit theory predicts that the sequence of means approaches infinity. But what about the sequence of medians? The limit of this sequence, if it exists, could be used to define a "fair" fee which would give Peter approximately a 50% chance of finishing a 2048-game series of Petersburg games with positive cumulative winnings.

The Limiting Median. If an n-game series of Petersburg games, having an average winning of $A(n)$ crowns per game, is repeated r times, a sequence $A_1(n), A_2(n), \ldots, A_r(n)$ of average winnings per game is gener-

---

[1] Coin tosses were simulated using the BASIC random number generator, RND, as a seed in the HP-25 generator, $U_{i+1} = \text{FRAC}(U_i + \pi)^5$. This was done since RND produced a distribution of game lengths which badly failed a chi-square goodness of fit test.

ated. Let $M_{n;r}$ denote the *median** of these r numbers. For example, our repetitions of Buffon's experiment game $M_{2048;1000} = 6.84$.

The possible average winnings for an n-game series are clearly discrete with no two averages closer than $1/n$. Theoretically, the probability of the occurrence of each possible average can be calculated. An average winning of $(n + 1)/n$ crowns per game, for example, can occur only if one game ends on the second toss and all other games end on the first toss. Since this can happen in n different ways, the probability of this average winning is

$$\left(\frac{1}{4}\right) \cdot \left(\frac{1}{2}\right)^{n-1} \cdot n = \frac{n}{2^{n+1}}.$$

Now let the possible values for the random variable $A(n)$ be written as an ordered sequence $a_1 < a_2 < a_3 < \cdots$. Then there must be a value a_i that satisfies either (i) or (ii):

(i) $P(A(n) < a_i) < \frac{1}{2}$ and $P(A(n) \leqslant a_i) > \frac{1}{2}$,

(ii) $P(A(n) \leqslant a_i) = \frac{1}{2}$.

In other words, there is a value a_i (in the ordered sequence of values for $A(n)$) at which the cumulative probability equals or exceeds $\frac{1}{2}$.

The behavior of $M_{n;r}$ as $r \to \infty$ is different in these two cases, as the following analysis indicates.

Case i. With probability one,[†] $M_{n;r} \to a_i$ as $r \to \infty$. The case $n = 4$ will serve as an illustration. The possible total four-game winnings are $4, 5, 6, 7, 8, 9, \ldots$. (In general, all consecutive integral totals between n and $2n$ are possible, but not all integral totals greater than $2n$ are possible.) The four-game average winnings that are possible are $1, 1\frac{1}{4}, 1\frac{1}{2}, 1\frac{3}{4}, 2, 2\frac{1}{4}, \ldots$. A routine calculation shows that $P(A(4) < 2\frac{1}{4}) = .472$, while $P(A(4) \leqslant 2\frac{1}{4}) = .520$. If a large number r of repetitions of this four-game series were performed, roughly 47.2% of the values in the sequence $A_1(4), A_2(4), \ldots, A_r(4)$ would be less than $2\frac{1}{4}$, roughly $52\% - 47.2\% = 4.8\%$ of the values would be equal to $2\frac{1}{4}$, and remaining 48% of the values would be greater than $2\frac{1}{4}$. The median, $M_{4;r}$, for these r values, would then be $2\frac{1}{4}$.

Of course, it is possible that due to random fluctuations, the observed frequencies for the values of $A(4)$ deviate from the above probabilities, but the deviations should be small if r is large. Thus, a limit statement "$\lim_{r \to \infty} M_{4;r} = 2\frac{1}{4}$" is required.

---

*Recall that the median of an ordered set of values $\{x_1 \leqslant x_2 \leqslant \cdots \leqslant x_r\}$ is any number M that satisfies $\#\{x_i < M\} \leqslant (r/2) \leqslant \#\{x_i \leqslant M\}$. Thus, $M = 3$ for $\{2, 3, 7\}$ and $M \in [3, 7]$ for $\{2, 3, 7, 8\}$. In general, $M = x_{(r+1)/2}$ if r is odd, and M can be taken as any value in the interval $[x_{r/2}, x_{(r/2)+1}]$ if r is even.

[†]The precise statement is that $P(\lim_{r \to \infty} M_{n;r} = a_i) = 1$ (i.e., the event that the random variable $M_{n;r}$ converges to a_i is an event whose probability is 1). In what follows, we will sometimes omit the phrase "with probability one," but it should be kept in mind that statements about random variables must be interpreted in this sense.

A formal proof that $\lim_{r\to\infty} M_{4;r} = 2\frac{1}{4}$ requires the law of large numbers. Let P_r, Q_r, and S_r denote the proportions of four-game averages in r repetitions that are respectively less than, equal to, or greater than, $2\frac{1}{4}$. The *Strong Law of Large Numbers* for proportions says that

$$P\left(\lim_{r\to\infty} P_r = .472, \ \lim_{r\to\infty} Q_r = .048, \ \lim_{r\to\infty} S_r = .480 \right) = 1.$$

The event that $\lim_{r\to\infty} P_r = .472$, $\lim_{r\to\infty} Q_r = .048$, $\lim_{r\to\infty} S_r = .480$ implies the event that $\lim_{r\to\infty} M_{4;r} = 2\frac{1}{4}$. Hence this last event also has probability equal to 1.

The reader should be able to extend this proof to an arbitrary value of n.

Technically, we cannot say that an entrance fee of $2\frac{1}{4}$ crowns per game gives Peter a 50% chance of winning a four-game series because of the probability of the series ending in a draw. In this case, the probability of a draw $P(A(4) = 2\frac{1}{4})$ is 4.8%, and the probability of positive cumulative winnings $P(A(4) > 2\frac{1}{4})$ is 48%. Since the probability of a draw becomes negligible with large values of n, we will consider the limiting median to be the "fair" fee for which we are looking.

Case ii.   Suppose $P(A(n) \leqslant a_i) = \frac{1}{2}$. Then $M_{n;r}$ oscillates between a_i and a_{i+1} as $r \to \infty$. The precise statement is that

$$P\left(\lim_{r\to\infty} \inf M_{n;r} = a_i \quad \text{and} \quad \lim_{r\to\infty} \sup M_{n;r} = a_{i+1} \right) = 1.$$

To see why this happens, consider the case $n = 2$. The possible two-game average winnings are $1, 1\frac{1}{2}, 2, 2\frac{1}{2}, \ldots$. A routine calculation gives $P(A(2) \leqslant 1\frac{1}{2}) = \frac{1}{2}$. If this two-game sequence were repeated a large number r of times, approximately 50% of the games would have average winnings of 1 or $1\frac{1}{2}$, and approximately 50% would have average winnings of 2 or more. If 50% or more of these average winnings were 1 or $1\frac{1}{2}$, then $M_{2;r}$ would be equal to $1\frac{1}{2}$. This follows by the definition of $M_{2;r}$ since $P(A(2) < \frac{3}{2}) = .25$. If slightly *less* than 50% (actually 43.75% to 50% since $P(A(2) = 2) = .0625$) were 1 or $1\frac{1}{2}$, then $M_{2;r}$ would be 2. The oscillation of $M_{2;r}$ is due to the fact that, with probability one, as $r \to \infty$ there will be infinitely many values of r for which the percentage is greater than 50%, and infinitely many values of r for which the percentage is slightly less than 50%. This is a fluctuation result from the law of large numbers. (It is similar to repeated tosses of a coin: the probability is one that the excess of heads over tails after r tosses will be positive for infinitely many values of r, and will also be negative for infinitely many values of r.)

We performed the necessary computation to obtain the limiting medians for $n \leqslant 8$. The results are given in Table 1. For $n > 8$, the computation involved in examining every case was prohibitive and since no general pattern was evident, we turned to the computer in search of a Monte Carlo solution to the problem.

Table 1.

n	Limiting Medians of the Sequence $\{M_{n;r}\}$
1	Oscillates between 1 and 2
2	Oscillates between $1\frac{1}{2}$ and 2
3	2
4	$2\frac{1}{4}$
5	$2\frac{2}{5}$
6	$2\frac{1}{2}$
7	$2\frac{5}{7}$
8	$2\frac{3}{4}$

Several computer runs were made to obtain sample values of $M_{n;r}$. We included values for $n \leqslant 8$ and $r = 10,000$ for comparison with the calculated limiting medians. The results of all runs are given in Table 2.

Table 2.

n	r	$M_{n;r}$	Limiting Median $\{M_{n;r}\}$
3	10,000	2	2
4	10,000	$2\frac{1}{4}$	$2\frac{1}{4}$
5	10,000	$2\frac{3}{5}$	$2\frac{2}{5}$
6	10,000	$2\frac{2}{3}$	$2\frac{1}{2}$
7	10,000	$2\frac{5}{7}$	$2\frac{5}{7}$
8	10,000	$2\frac{3}{4}$	$2\frac{3}{4}$
10	10,000	3.00	
12	10,000	3.08	
16	10,000	3.31	
64	10,000	4.28	
100	10,000	4.57	
256	10,000	5.36	
512	1,000	5.73	
1,024	1,000	6.10	
2,048	1,000	6.84	

The discrepancies between columns 3 and 4, for $n = 5$ and $n = 6$, may seem large, but remember the possible values for $M_{n;r}$ are discrete and differ by at least $1/n$. Furthermore, the calculated probabilities which gave rise to the limiting medians $2\frac{2}{5}$ and $2\frac{1}{2}$ in column 4 were .503 and .504, respectively. Thus, it was not surprising that, for large values of r, the sample medians $M_{n;r}$ were still oscillating between the values given in columns 3 and 4 of the table, falling after 10,000 repetitions in the least probable position.

In Search of a Fair-Fee Function. A function $f(n)$ which gives a good fit to the data of Table 2 was suggested to us by De Morgan's account of his student's 2048-game duplication of Buffon's empirical trial ([2], 185).

Table 3 is a reproduction from De Morgan's account in which he compares the distribution of the 2048 games of his student's trial with that of Buffon's. The fourth column of the table gives the numbers which he states "the theory asserts to be most probable." One is tempted to make the following intuitive analysis: There is a game in the fourth column that ends beyond the 11th toss. If this game actually ends on the 12th toss, the average winning per game for the 2048-game series is $(1024 \cdot 1 + 512 \cdot 2 + \cdots + 1 \cdot 2^{11} + 1 \cdot 2^{12})/2048 = 6.5$ crowns per game. If this game ends on the 13th toss, the average winning is $(1024 \cdot 1 + 512 \cdot 2 + \cdots + 1 \cdot 2^{11} + 1 \cdot 2^{13})/2048 = 7.5$ crowns per game. Note that our sample median $M_{2048;1000} = 6.84$ fell between these two values.

Table 3.

Throw on which first head appears	Frequency of occurrence in 2048-game series		
	Buffon	De Morgan's student	Most probable
1st	1061	1048	1024
2nd	494	507	512
3rd	232	248	256
4th	137	99	128
5th	56	71	64
6th	29	38	32
7th	25	17	16
8th	8	9	8
9th	6	5	4
10th	0	3	2
11th	0	1	1
12th	0	0	
13th	0	0	
14th	0	1	1
15th	0	0	
16th	0	1	
Beyond 16th	0	0	

Generalizing to an n-game series, where $n = 2^m$, we might further conjecture that for large r, the median $M_{n;r}$ should fall between the bounds of

$$(2^{m-1} \cdot 1 + 2^{m-2} \cdot 2 + \cdots + 1 \cdot 2^{m-1} + 1 \cdot 2^m)/2^m = \frac{m}{2} + 1$$

and

$$(2^{m-1} \cdot 1 + 2^{m-2} \cdot 2 + \cdots + 1 \cdot 2^{m-1} + 1 \cdot 2^{m+1})/2^m = \frac{m}{2} + 2.$$

Since (in this case) $m = (\log n / \log 2)$, the conjecture for an arbitrary n, not necessarily of the form 2^m, is that $M_{n;r}$ will fall between the bounds of

$$\frac{\log n}{2 \log 2} + 1 \quad \text{and} \quad \frac{\log n}{2 \log 2} + 2.$$

All calculated and computer-generated medians given in Table 1 and Table 2 fall between these bounds, with the limiting median of $\{M_{1;r}\}$ actually oscillating between the two bounds.

One might hope to find a value k such that the limiting median of $\{M_{n;r}\}$, when it exists, would equal $(\log n / 2 \log 2) + k$. This, however, was found to be impossible since different values of k ranging from 1.21 to 1.31 would be necessary to fit the calculated limiting medians in Table 1. For $n > 8$, however, the functions

$$F(n) = \frac{\log n}{2 \log 2} + 1.3 \tag{1}$$

fit the computer-generated sample medians surprisingly well. (See Table 4.)

The linear least squares fit of $M_{n;r}$ to $\log n$, with the data of Table 4, is $M_{n;r} = 1.63 \log n + 1.34$. If, on the other hand, we fit $M_{n;r}$ to

$$\frac{\log n}{2 \log 2} + k = 1.66 \log n + k,$$

the best estimate of k is 1.28.

Table 4.

n	r	$M_{n;r}$	$\dfrac{\log n}{2 \log 2} + 1.3$	$M_{n;r} - \left(\dfrac{\log n}{2 \log 2} + 1.3 \right)$
9	10,000	2.89	2.88	.01
10	10,000	3.00	2.96	.04
12	10,000	3.08	3.09	−.01
16	10,000	3.31	3.30	.01
64	10,000	4.28	4.30	−.02
100	10,000	4.57	4.62	−.05
256	10,000	5.36	5.30	.06
512	1,000	5.73	5.80	−.07
1024	1,000	6.10	6.3	−.20
2048	1,000	6.84	6.8	.04

A comparison of equation (1) with Feller's result ([3], 236–237) is interesting. He defines a fee f_n, for an n-game series of Petersburg games, to be "fair" if for every $\epsilon > 0$:

$$P \left\{ \left| \frac{A(n)}{f_n} - 1 \right| > \epsilon \right\} \to 0, \tag{2}$$

where $A(n)$ is the average per-game winning for the n-game series. Feller then proves that $f_n = (\log n / 2 \log 2)$ produces a fair game. Although Feller's definition of fair is entirely different than ours, our fair fee given by equation (1) would also be Feller-fair. In fact, it is easily seen from definition (2) that if $f_n = (\log n / 2 \log 2)$ is Feller-fair, so also is

$$\frac{\log n}{2 \log 2} + k$$

for any value of k. Thus, Feller's result is of little practical value to a gambler.

The Solution. If we neglect the probability of a draw, the limiting median of the sequence $\{M_{n;r}\}$ gives the entrance fee which would give Peter a 50% chance of winning an n-game series of Petersburg games. For $n \leqslant 8$, we have computed and shown this fee to be between

$$\frac{\log n}{2 \log 2} + 1 \quad \text{and} \quad \frac{\log n}{2 \log 2} + 2.$$

For $n \geqslant 8$, our Monte Carlo results indicate that this fee is approximately $F(n) = (\log n / 2 \log 2) + 1.3$.

Rationalizing the Paradox. The predicted median in (1) of the average winnings of an n-game series depends logarithmically on n and approaches infinity with increasing n. This fact gives insight into the sense in which the mathematical expectation of a single game is infinite, and why Buffon's 2048-game trial is harmonious with the mathematical theory.

The paradox of the Petersburg game lies primarily in the use of the word "expectation." Suppose Peter plays Paul a game of chance in which Peter has one chance in one thousand to win ten thousand crowns. Although Peter's mathematical expectation is 10 crowns, his chance of winning is so small that he realistically expects nothing. However, if he could play millions of games, he would expect an average winning of approximately 10 crowns per game. Similar reasoning applies to the Petersburg game. Although the mathematical expectation is infinite, the number of games played would have to approach infinity before one could realistically expect average winnings per game to approach infinity. Taking $F(n) = 12$, in equation (1), for example, we see that one would have to play about 2,000,000 games in order to expect with probability 50% an average winning per game exceeding a modest 12 crowns.

We simulated a 2,000,000-game series a single time ($r = 1$), printing out the cumulative average winnings per game in increments of 10,000. The result is graphed in Figure 2. Note how the average tends to drift down until a long run of tails occurs which kicks it up. The big jump between 730,000 and 740,000 games was caused by a run of 21 tails. The dotted line in Figure 2 shows the conjectured median average $(\log n / 2 \log 2) + 1.3$. The agreement between the average per-game winnings and the conjectured median average (especially for $n > 1,000,000$) seemed somewhat surprising since this was a single trial ($r = 1$); the dotted line describes the conjectured "limiting value" of the sequence $M_{n;r}$ ($r = 1, 2, \ldots$) for each fixed n.

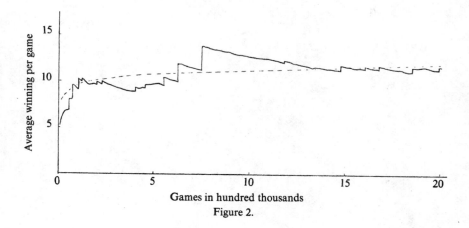

Figure 2.

Conclusion and Challenge. Our results indicate that Buffon's famous empirical trial of the 2048 Petersburg games, yielding an average winning of 4.9 crowns per game, was slightly lower than one would expect in the sense that if Buffon's experiment were repeated many times, the average would exceed 6.8 crowns approximately 50% of the time.

We have given an approximation to the median average winnings for an n-game series, and we have compared this result with many computer-simulated series, including one series of two million games. These results give the modern student valuable insight, not available to Buffon and his contemporaries, into the Petersburg Paradox.

There is little mathematical theory available for the analysis of distributions with infinite means such as the Petersburg game. Perhaps our empirical results will point the way for the development of additional theory. In particular, we challenge the interested reader to prove (or disprove) our conjecture that the limiting median of $\{M_{n;r}\}$ falls between

$$\frac{\log n}{2\log 2} + 1 \quad \text{and} \quad \frac{\log n}{2\log 2} + 2$$

for all n.

REFERENCES

1. C. Boyer, A History of Mathematics, Wiley, New York, 1968.
2. A. De Morgan, Formal Logic or the Calculus of Inference Necessary and Probable, Taylor and Walton, 1847.
3. W. Feller, An Introduction to Probability Theory and Its Applications, vol. 1, 2nd ed., Wiley, New York, 1957.
4. I. Todhunter, A History of the Mathematical Theory of Probability, Cambridge, 1965.

6. CALCULATORS & COMPUTERS

"To err is human, but when the eraser wears out ahead of the pencil, you're overdoing it."

—J. Jenkins, The Peter Prescription (1973)

"Seeing there is nothing (right well beloved Students in the Mathematickes) that is so troublesome to Mathematicall practise, nor that doth more molest and hinder Calculators, than the Multiplications, Divisions, square and cubical Extractions of great numbers, which besides the tedious expence of time are for the most part subject to many slippery errors. I began therefore to consider in my minde, by what certaine and ready Art I might remove those hindrances."

—John Napier (1614)

"Computers are fantastic: in a few moments they can make a mistake so great that it would take many men many months to equal it."

—M. Meacham, The Peter Prescription (1973), 50

"As time goes on, the computer appears to become increasingly intelligent. Once the programs are so complex that their inventors cannot quickly predict all possible responses, the machines will have the appearance of, if not intelligence, at least free will."

—Carl Sagan, The Dragons of Eden (1977), 220

"It's great to be great, but it's greater to be human.

—Will Rogers

Rapid change often cancels assumptions and constraints of the past only to create new ones upon which to base future decisions. Consider the new culture being created by the application of computer technology. What premises do we, as mathematics educators, adopt in the presence of this technology?

The decade of the Sixties brought forth a "new generation" of computers. The increased presence of minicomputers, the introduction to an interactive mode, and the powerful and simplified programming language BASIC all served to put the power of the computer into the hands of a much larger and younger group of users, including teachers of mathematics and their students.

The Seventies brought us the widespread availability of hand-held calculators. As their prices dropped, we found more and more students showing up in class with these devices. Toward the end of the decade, low-cost microcomputers entered the mass marketplace.

Expanded capabilities and applications that cost less and are easier to use are becoming increasingly available. How can the prospect of interactive graphics fail to spark the imagination of even the meanest (mean = average) math teacher?

Probably every computer programmer knows some mathematics, but does every mathematics teacher know some programming language? The computer-literate student who learns mathematics is likely to do more interesting things with the computer as a result of that training. (A knowledge of vectors, matrices, analytic geometry, calculus, and gradients, for example, would be valuable information for students interested in computer graphics.) Mathematics teachers with some computer literacy will also be able to do much more interesting things in mathematics and mathematics education than was formerly possible.

This change, in the form of computer technology, will raise many questions about why and what teachers do, and how it is to be done. There will be interesting challenges at every juncture in the changing curricular and pedagogical process: Have parts of the curriculum fallen into obsolescence? Are there innovative ways of doing some of our tasks that would enhance their impact? Should some parts of the curriculum be restructured or receive more emphasis than before? What new ideas, that were unthinkable without this technology, can now be explored?

The following papers address some of these questions. John Varner's "Direct Reduction Loans" demonstrates that the difficulty of calculating formulas associated with direct reduction loans need no longer be a problem for consumers with an electronic calculator.

In "Calculators and Polynomial Evaluation," J. F. Weaver adds another facet to polynomial evaluation by considering several types of calculator algorithms to facilitate the process.

Michael Burke's "Integrating Mathematics & Computer Science: A Case for LISP" provides an interesting introduction to the computer language

LISP, a mathematical language with attendant structure and formalism sufficient to solve a broader class of problems than can other programming languages. LISP can also be used to check and prove the consistency of programs written in other languages as well as in formal axiomatic sets.

In "Simulating Sampling from Normal Populations," Catherine Lilly demonstrates the power of the computer in carrying out a large number of routine calculations. Here the computer is used to produce a sequence of values that are normally distributed (as predicted by the Central Limit Theorem), thereby simulating a sample from a normal population.

Anyone who uses a computer realizes that there are limitations on the subset of the rational numbers it can deal with. There is a finite greatest number, and a somewhat lesser degree of precision available for most of the numbers (i.e., 6, 8, 12, 20, etc., displaying digits). Gerard Kiernan's "Extended Precision Arithmetic—A Numeric Approach" offers a program to extend the precision of some computations.

Direct Reduction Loans

John T. Varner III, *University of South Carolina, Sumter, SC*

In most books titled Business Mathematics, if there is a treatment of direct reduction loans, it is usually restricted to the definition of what is meant by such a loan. The mathematics is not done and all problems are solved by using tables.

It seems to me that this is an ideal opportunity to present some applied mathematics at an elementary level which is useful to anyone who is paying on a home mortgage, an automobile, or obtaining a direct reduction loan for any purpose. Moreover, the formulas that are developed require the calculation of $(1 + i)^N$, where N is an integer and $0 < i < 1$. This is an excellent opportunity to use an electronic calculator. In fact, one reason textbooks have avoided the mathematics and approached the topic through tables is because of the difficulty of calculating $(1 + i)^N$.

For the reader who is not sure about what is meant by a direct reduction loan, it is simply a loan on which interest is charged on the unpaid balance. Most credit card loans or purchases are a form of direct reduction loan.

Suppose a loan of A dollars is to be repaid in N equal installments of M dollars with the percentage rate per period equal i. Let $A_k = $ loan balance after payment k. The following derivation of the formula relating A, M and i seems more direct than the usual one seen in textbooks.

$$A_i = A + Ai - M = A(1 + i) - M,$$

$$A_2 = A_1 + A_1 i - M = A_1(1 + i) - M = A(1 + i)^2 - M[(1 + i) + 1],$$

$$A_3 = A_2 + A_2 i - M = A_2(1 + i) - M$$

$$= A(1 + i)^3 - M[(1 + i)^2 + (1 + i) + 1].$$

By mathematical induction,

$$A_k = A(1 + i)^k - M[(1 + i)^{k-1} + (1 + i)^{k-2} + \cdots + (1 + i) + 1]. \quad (1)$$

Using the formula for the sum of a geometric sequence with $r = 1 + i$, (1) becomes

$$A_k = A(1 + i)^k - M \frac{[(1 + i)^k - 1]}{i}. \quad (2)$$

If the loan is to be paid off after N payments, A_N must equal zero. Thus, letting $k = N$ and solving $A_N = 0$ in (2) gives

$$M = \frac{Ai(1 + i)^N}{(1 + i)^N - 1} \tag{3}$$

and

$$A = \frac{M\left[(1 + i)^N - 1\right]}{(1 + i)^N \cdot i}. \tag{4}$$

According to Regulation Z of The Consumer Credit Protection Act, most installment loan contracts must be in terms of annual percentage rate. If the annual percentage rate is r, then the monthly payment can be computed from (3) by taking $i = r \div 12$.

For example, if \$4,200 is financed on a new automobile for three years at an annual percentage rate of twelve, then we have $N = 36$ and $i = .12/12 = .01$. The monthly payment, computed from (3), is

$$M = \frac{(4200)(.01)(1.01)^{36}}{(1.01)^{36} - 1} = 139.50.$$

As another example, consider the monthly mortgage payment on a \$40,000 house for thirty years at 9%. We have $N = 360$; $i = .09/12 = .0075$, and $M = \$321.85$. Note that the cost of credit is $MN - A = \$75,866$ (considerably more than the cost of the house).

To compute the balance on this mortgage after payment 240, use formula (2) with $k = 240$. This gives $A_{240} = \$25,406.67$.

It doesn't take many examples to convince the student that credit is costly.

Calculators and Polynomial Evaluation

J. F. Weaver, *The University of Wisconsin, Madison, WI*

The intent of this paper is to suggest and illustrate how electronic hand-held calculators, especially nonprogrammable ones with limited data-storage capacity, can be used to advantage by students in one particular aspect of work with polynomial functions.

It is well known that calculators can facilitate greatly the evaluation of polynomial functions (especially when working with nonintegral values), along with sketching graphs of such functions and "closing in on" their real zeros (when such exist). But the facet of polynomial evaluation to be emphasized in this paper seems to have received relatively little attention with respect to calculator use.

First it may be well to summarize a bit of mathematical background that is assumed to be familiar to students as the basis upon which calculator application is built.* Since calculators operate on (and otherwise process) only *rational* numbers and proper subsets thereof, or rational approximations to real numbers, the set of rationals will be the universe for the following summary and subsequent material (unless explicitly stated otherwise).

Polynomials, Quotients and Remainders

Consider a polynomial function of the form

$$f(x) = c_n x^n + c_{n-1} x^{n-1} + c_{n-2} x^{n-2} + \cdots + c_2 x^2 + c_1 x + c_0, \quad (1)$$

where n is a positive integer.

Dividing $f(x)$ by $x - a$ generates a quotient q and a remainder r such that

$$f(x) = (x - a)q + r.$$

---

*In some lower level precalculus courses this mathematical content may be *developed* as part of the course work rather than "assumed." In such instances that development might be *facilitated* by suitable adaptation of the materials and suggestions presented in this paper.

Here q may be expressed in the form

$$k_{n-1}x^{n-1} + k_{n-2}x^{n-2} + k_{n-3}x^{n-3} + \cdots + k_1x + k_0, \qquad (1.1)$$

where $k_{n-1} = c_n$.

Using synthetic division, we may show our work in the following abbreviated way

a	c_n	c_{n-1}	c_{n-2}	$\cdots$	c_2	c_1	c_0
		$k_{n-1}a$	$k_{n-2}a$	$\cdots$	k_2a	k_1a	k_0a
	k_{n-1}	k_{n-2}	k_{n-3}	$\cdots$	k_1	k_0	r

where

$$k_{n-1} = c_n, \qquad (1.11)$$

$$k_{n-2} = c_na + c_{n-1}, \qquad (1.12)$$

$$k_{n-3} = (c_na + c_{n-1})a + c_{n-2}, \qquad (1.13)$$

$$\cdots\cdots\cdots\cdots\cdots\cdots\cdots$$

$$k_1 = (((c_na + c_{n-1})a + c_{n-2})a + \cdots + c_3)a + c_2, \qquad (1.14)$$

$$k_0 = (((c_na + c_{n-1})a + c_{n-2})a + \cdots + c_2)a + c_1, \qquad (1.15)$$

and

$$r = ((((c_na + c_{n-1})a + c_{n-2})a + \cdots + c_2)a + c_1)a + c_0. \qquad (1.2)$$

This same information (the terms of q and r when $f(x)$ is divided by $x - a$) may be generated by first expressing $f(x)$ in a "nested parentheses" form,

$$f(x) = ((((c_nx + c_{n-1})x + c_{n-2})x + \cdots + c_2)x + c_1)x + c_0 \qquad (2)$$

and then computing

$$f(a) = ((((c_na + c_{n-1})a + c_{n-2})a + \cdots + c_2)a + c_1)a + c_0.$$

Terms of quotient, q:

Coefficient of	x^{n-1}:	$k_{n-1} = c_n$	(2.11)
Coefficient of	x^{n-2}:	$(k_{n-2}\leftarrow$	(2.12)
Coefficient of	x^{n-3}:	$((\leftarrow\quad k_{n-3}\quad\longrightarrow$	(2.13)
$\cdots\cdots\cdots\cdots\cdots\cdots\cdots\cdots\cdots$			
Coefficient of	x:	$(((\leftarrow\quad k_1\longrightarrow$	(2.14)
Constant term:		$((((\leftarrow\quad k_0\longrightarrow$	(2.15)
And the remainder:		$(((((\leftarrow\quad r\longrightarrow$	(2.2)

Evaluating Polynomials

When we "*evaluate a polynomial*" $f(x)$ by computing $f(a)$, we in effect compute the *remainder* r when $f(x)$ is divided by $x - a$.* (In the special

---

*The *remainder theorem* (and subsequently the *factor theorem*) can be made explicit rather than informally implied, if desired.

case of $r = 0$, it is clear that $x - a$ is a factor of $f(x)$, and a is a root of $f(x) = 0$.)

A polynomial may be evaluated by using either form (1) or form (2) of $f(x)$ as the basis for computing each $f(a)$ desired. However, if there is interest in q as well as r, form (2) clearly is the preferred one to use.

Furthermore, if we wish to effect "*calculator*-assisted" polynomial evaluation(s), form (2) has a distinct advantage over form (1) in that $f(a)$ may be computed by using only the first power of a. If computation of some $f(a)$ based upon form (1) is to be executed efficiently, for $n > 2$ it is essential that a calculator have a $\boxed{y^x}$ key. However, many calculator models will display some form of error message if that key is used in conjunction with any $a < 0$, which is an intolerable situation.

Thus, further consideration of polynomial evaluation(s) in this paper will emphasize the importance of form (2) as the basis for such computation(s). A preliminary example may serve to illustrate the preceding points and several others that will have a bearing upon the work to follow.

Consider

$$f(x) = 2x^5 - x^4 + 35x^2 - 69x + 45, \tag{3}$$

which may be rewritten in the following manner to conform to the format of (1):

$$f(x) = 2x^5 + (-1)x^4 + 0x^3 + 35x^2 + (-69)x + 45. \tag{3.1}$$

To calculate $f(2)$, for instance, it is sufficient to use (3) rather than (3.1):

$$f(2) = 2(2^5) - 2^4 + 35(2^2) - 69(2) + 45$$

$$= 2(32) - 16 + 35(4) - 69(2) + 45$$

$$= 64 - 16 + 140 - 138 + 45$$

$$= 95. \tag{3.2}$$

That is, when $f(x)$ is divided by $x - 2$, the *remainder* is 95, and so $x - 2$ is not a factor of $f(x)$. But no information is given regarding the terms of the quotient q.

Similarly,

$$f(-3) = 2(-3)^5 - (-3)^4 + 35(-3)^2 - 69(-3) + 45$$

$$= 2(-243) - 81 + 35(9) - 69(-3) + 45$$

$$= -486 - 81 + 315 + 207 + 45$$

$$= 0. \tag{3.3}$$

Thus, when $f(x)$ is divided by $x - (-3) = x + 3$, we obtain $r = 0$. Therefore, $x + 3$ is a factor of $f(x)$, and -3 is a root of $f(x) = 0$. But, as with (3.2), no information is revealed regarding the terms of q.

Although the x^3 term with its 0 coefficient was simply disregarded in the preceding evaluations, that cannot be done if evaluations are generated by synthetic division based upon a nested-parentheses format such as (2) for $f(x)$. For instance, let us rewrite (3.1) as

$$f(x) = ((((2x + -1)x + 0)x + 35)x + -69)x + 45. \qquad (3.4)$$

Although it is not *necessary* to express $f(x)$ in a nested-parentheses form to *do* synthetic division, such a form is helpful in *conceptualizing* the procedure —which in the case of $f(2)$, for instance, may be shown in a compact algorithm such as

2	2	−1	0	35	−69	45
		4	6	12	94	50
	2	3	6	47	25	95.

As before, $r = 95$. But we also can write the terms of q, which may be viewed as the depressed polynomial $g(x) = 2x^4 + 3x^3 + 6x^2 + 47x + 25$.

(If students are relatively inexperienced or somewhat insecure with the synthetic division procedure, a mnemonic form such as Table 1 may be used, which is completed as Table 2 for $f(2)$.)

Table 1.

$f(x) =$									
x	c_8	c_7	c_6	c_5	c_4	c_3	c_2	c_1	c_0
$\times$									
$+$									
	k_7	k_6	k_5	k_4	k_3	k_2	k_1	k_0	r

Table 2.

$f(x) = 2x^5 - x^4 + 35x^2 - 69x + 45$									
x	c_8	c_7	c_6	c_5	c_4	c_3	c_2	c_1	c_0
2				2	−1	0	35	−69	45
$\times$					4	6	12	−94	50
$+$				2	3	6	47	25	95
	k_7	k_6	k_5	k_4	k_3	k_2	k_1	k_0	r

Finally, in connection with this background summary, $f(-3)$ has been computed using a compact synthetic division algorithm

$$
\begin{array}{r|rrrrrr}
-3 & 2 & -1 & 0 & 35 & -69 & 45 \\
& & -6 & 21 & -63 & 84 & -45 \\
\hline
& 2 & -7 & 21 & -28 & 15 & 0.
\end{array}
$$

Table 3 is a "helper" version of the same algorithm:

Table 3.

					$f(x) = 2x - x^4 + 35x^2 - 69x + 45$				
x	c_8	c_7	c_6	c_5	c_4	c_3	c_2	c_1	c_0
-3			2	-1	0	35	-69	45	
$\times$					-6	21	-63	84	-45
$+$				2	-7	21	-28	15	0
	k_7	k_6	k_5	k_4	k_3	k_2	k_1	k_0	r

As before, $r = 0$ establishes that $x + 3$ is a factor of $f(x)$, and -3 is a root of the equation $f(x) = 0$. It also is known that $q = g(x) = 2x^4 - 7x^3 + 21x^2 - 28x + 15$, and further work may be done with that function or with the depressed equation $g(x) = 0$.

Calculator-assisted Polynomial Evaluation

How can a calculator be used to evaluate a particular polynomial function so that along with r, the terms of q or $g(x)$ also are generated in order when $f(x)$ is (in effect) divided by $x - a$?

Figure 1 suggests an iterative algorithm for doing this—an algorithm consistent with the notation used earlier in this paper and relatively independent of calculator type, assuming (at least) one data-storage register ("memory") or its equivalent. Such an algorithm (Figure 1 or a similar one) clearly should not be presented to students as is done here, as a *fait accompli*, but should be developed by or with them based upon a nested parentheses form for $f(x)$ coupled with the synthetic division procedure.

Several things should be noted about Figure 1:
  (a) It is assumed that the algorithm will be applied to polynomial functions for which $n > 1$.
  (b) Steps such as 40, 60, 70 and 90 are exclusively "mental," and are to be executed wholly independently of a calculator.

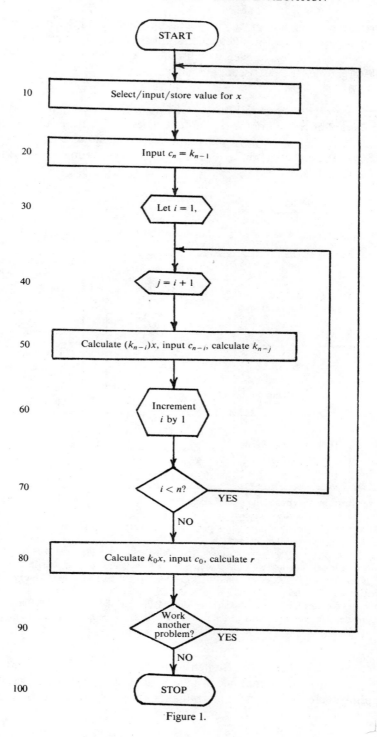

Figure 1.

(c)   The specific way in which processing is effected in step 20 or 50 or 80 is *not* calculator-independent and may vary from one calculator type to another, or even within the same calculator type. This leads directly to the next section of this paper.

Illustrative calculator algorithms.   The algorithms to be presented illustrate a variety of calculator types, characteristics, and features that in many cases are independent of particular brand-name calculator models. Table 4 summarizes the essential similarities and differences among the eight algorithms to be detailed.

Table 4.

Calculator algorithm	What "logic" or data entry/processing system is used?	What kind of "memory" or other means is used to store values of x?	What other calculator or algorithm features should be noted?
A		4-key "accumulating":	No "chaining" used
B	Algebraic, with no operational hierarchy	MC M + M − MR	"Chaining" is used
C			No "chaining" used
D		2-key: STO RCL	"Chaining" is used
E	Algebraic, with a multiply-before-add operational hierarchy*	2-key: STO RCL	No parentheses keys are used
F			Nested-parentheses keys are used[†]
G	RPN (reverse Polish notation)	2-key: STO RCL	
H		3- or 4-level operational stack[‡]	

*Texas Instruments applies the trademarked label AOS (Algebraic Operating System) to its calculator models that utilize this hierarchy.

[†]In order to use this algorithm with a particular calculator, it should be capable of nesting as many as $n - 1$ levels of parentheses.

[‡]It also must be true (as is the case with Hewlett-Packard's RPN models) that when the stack "drops" in conjunction with an executed binary operation, the number in the T (top) register is *retained in* T (i.e., copied into itself) as well as stored in the next "lower" register immediately adjacent to the T (e.g., Z in a 4-level stack; Y in a 3-level stack). General applicability of Algorithm H is *invalidated* for those RPN calculators (e.g., National Semiconductor's models) in which the T register is *cleared* (i.e., *set to zero*) in conjunction with the execution of a binary operation on the numbers in the Y and X registers.

The algorithms are not presented as calculator *programs*, but rather as keystroke sequences or routines to be executed *manually*. Along with each algorithm there is included an indication of the number of manual keystrokes (excluding those for data input) needed to execute that algorithm.

Only *one* "memory" or independent data-storage register is assumed in connection with any calculator, so *no address label* has been indicated when access to that register is required in an algorithm. (Keystroking would need to be modified accordingly if a calculator had more than one "memory.")

As was true for Figure 1, each of the eight calculator algorithms that follows is presented here as a *fait accompli*. From an instructional standpoint, however, each should emerge from work by or with students using their calculators.

Calculator Algorithm *A*.

STEP	INSTRUCTION	INPUT	KEY(S)			OUTPUT
1		x	MC	M +	×	
		c_n	+			
		c_{n-1}	=			k_{n-2}
2	Iterate $n-1$ times for		×	MR	+	
	$i=2$ thru $i=n$	c_{n-i}	=			k_{n-3} thru r
3	Return to step 1 for next x					
			$4n+1$ keystrokes			

Calculator Algorithm *B*.

STEP	INSTRUCTION	INPUT	KEY(S)			OUTPUT
1		x	MC	M +	×	
		c_n	+			
		c_{n-1}	×			k_{n-2}
2	Iterate $n-2$ times for		MR	+		
	$i=2$ thru $i=n-1$	c_{n-i}	×			k_{n-3} thru k_0
3			MR	+		
		c_0	=			r
4	Return to step 1 for next x					
			$3n+2$ keystrokes			

Calculator Algorithm C.

STEP	INSTRUCTION	INPUT	KEY(S)			OUTPUT
1		x	STO	$\times$		
		c_n	+			
		c_{n-1}	=			k_{n-2}
2	Iterate $n-1$ times for		$\times$	RCL	+	
	$i=2$ thru $i=n$	c_{n-i}	=			k_{n-3} thru r
3	Return to step 1 for next x					
			$4n$ keystrokes			

Calculator Algorithm D.

STEP	INSTRUCTION	INPUT	KEY(S)			OUTPUT
1		x	STO	$\times$		
		c_n	+			
		c_{n-1}	$\times$			k_{n-2}
2	Iterate $n-2$ times for		RCL	+		
	$i=2$ thru $i=n-1$	c_{n-i}	$\times$			k_{n-3} thru k_0
3			RCL	+		
		c_0	=			r
4	Return to step 1 for next x					
			$3n+1$ keystrokes			

Calculator Algorithm E.

STEP	INSTRUCTION	INPUT	KEY(S)			OUTPUT
1		x	STO	$\times$		
		c_n	+			
		c_{n-1}	=			k_{n-2}
2	Iterate $n-1$ times for		$\times$	RCL	+	
	$i=2$ thru $i=n$	c_{n-i}	=			k_{n-3} thru r
3	Return to step 1 for next x					
			$4n$ keystrokes			

Calculator Algorithm F.

STEP	INSTRUCTION	INPUT	KEY(S)				OUTPUT
1	Iterate $n-1$ times		(				
2		x	STO	$\times$			
		c_n	+				
		c_{n-1}	)				k_{n-2}
3	Iterate $n-2$ times for		$\times$	RCL	+		
	$i=2$ thru $i=n-1$	c_{n-i}	)				k_{n-3} thru k_0
4			$\times$	RCL	+		
		c_0	=				r
5	Return to step 1 for next x						
			$5n-1$ keystrokes				

Calculator Algorithm G.

STEP	INSTRUCTION	INPUT	KEY(S)			OUTPUT
1		x	STO			
		c_n	$\times$			
		c_{n-1}	+			k_{n-2}
2	Iterate $n-1$ times for		RCL	$\times$		
	$i=2$ thru $i=n$	c_{n-i}	+			k_{n-3} thru r
3	Return to step 1 for next x					
			$3n$ keystrokes			

Calculator Algorithm H.

STEP	INSTRUCTION	INPUT	KEY(S)			OUTPUT
1		x	ENTER↑	ENTER↑	*ENTER↑	
		c_n				
2	Iterate n times for		$\times$			
	$i=1$ thru $i=n$	c_{n-i}	+			k_{n-2} thru r
3	Return to step 1 for next x					
*	Omit for $n=2$					
			$2n+3$ keystrokes			

Output from any of the calculator algorithms may be recorded on a form such as Table 5, which is illustrated by Table 6.

Table 5.

$f(x) =$									
	c_8	c_7	c_6	c_5	c_4	c_3	c_2	c_1	c_0
x	k_7	k_6	k_5	k_4	k_3	k_2	k_1	k_0	r

Table 6.

$f(x) = 2x^5 - x^4 + 35x^2 - 69x + 45$									
	c_8	c_7	c_6	c_5	c_4	c_3	c_2	c_1	c_0
				2	-1	0	35	-69	45
x	k_7	k_6	k_5	k_4	k_3	k_2	k_1	k_0	r
2				2	3	6	47	25	95
-3				2	-7	21	-28	15	0

In Conclusion

As students use their calculators to develop and apply algorithms such as (or similar to) the ones illustrated in this paper, various questions profitably may be raised for consideration and discussion. For instance:

1. Which algorithm(s), if any, is(are) more "meaningful" than others in the sense of making the synthetic division process readily comprehensible?

2. Which algorithm(s) is(are) most efficient in the sense of requiring the fewest number of keystrokes to execute manually?

3. What other factor(s) is(are) of nontrivial consequence in judging the efficacy of calculator algorithms?

And for students having programmable calculators . . .

How readily can these manually-executed algorithms be modified or adapted or extended and incorporated into programs for calculators that have one or more features such as

looping capability,

a multiplicity of addressable registers for data storage,

indirect addressing,

printing capability,

etc.?

All in all, suitable use of calculators can enhance as well as facilitate numerous aspects of work with polynomial functions and evaluation. One such possibility has been suggested and illustrated in this paper.

Integrating Mathematics and Computer Science: A Case for LISP

Michael E. Burke, *San Jose State University, San Jose, CA*

Today's mathematics undergraduates are becoming more and more involved with computer science concepts and applications. Mathematics educators should therefore emphasize the thought process necessary to become proficient problem solvers. This paper offers some ideas of how to accomplish this goal without placing an intolerable burden of course work on the mathematics undergraduate.

A student's first acquaintance with computer science is usually an introductory course in programming. Such a course usually removes much of the mystery of computers, once students understand how computers are being used. After his first programming experience and the usual dose of calculus, the interested student is ushered into numerical analysis, linear programming, graph theory, combinatorics, etc. These courses are useful and important. They broaden students' knowledge of the applications of computers, and they enable students to more fully experience the theoretical aspects of mathematics. However, there are shortcomings in this approach.

First, the desire to understand more about computer science is strong among students of mathematics. Their needs are not being met. Those who take additional computer science courses sometimes find that these courses tend to be oriented to very specific problem areas. Therefore those who complete such courses still have a vague picture of the computer science landscape. Secondly, the programming language(s) and techniques they originally learned tend to govern their thought processes in the future. Thus, students schooled in, say FORTRAN, think problems through in FORTRAN—rather than looking at the abstract nature of the problem. Effective thinking in programming requires general inventiveness, not just more cleverness in the use of a particular language. Finally, math students are not aware of the interrelationship between mathematics and computer science. Specifically, there is more mathematics to computer science than that of finding algorithms to solve problems.

Students need to know something about recursion, symbolic programming, architecture, data structures, compilers, interpreters, and programming languages in general. They can get this background, as well as a better understanding of the role of mathematics in computer science, by taking a one-year course centered around the LISP programming language. The goal of such a course is not to teach students another language, but rather to provide them with the opportunity to develop general reasoning skills which can be usefully applied in various contexts.

For those unfamiliar with LISP, we will introduce the main control structure and give some examples from number theory. With LISP one can represent the integers, the arithmetic functions $+$, $-$ and $*$, and the logical operators (also called predicates) $=$, $<$, $>$, $\leqslant$, $\geqslant$, and $\neq$. Next we introduce the construct of a conditional expression. In LISP, this takes the form:

$$[\,p_1 \rightarrow e_1;\, p_2 \rightarrow e_2;\, \ldots;\, p_n \rightarrow e_n\,], \tag{1}$$

where $p_1, p_2, \ldots, p_n$ are predicates and $e_1, e_2, \ldots, e_n$ are LISP expressions. The value of the conditional expression is the value of e_i that corresponds to the first p_i which evaluates to true. Thus (1) has value e_1 if p_1 is true; (1) has value e_2 if p_1 is false and p_2 is true; etc. If all p_i evaluate to false, then (1) is undefined. And (1) is undefined if the value of e_i, corresponding to the first true p_i, is undefined.

For example we could define a function which computes the absolute value of a number as:

$$\mathrm{abs}[\,x\,] \Leftarrow [\,x < 0 \rightarrow -x;\, x \geqslant 0 \rightarrow x\,].$$

Here $\{x < 0\} = p_1 \rightarrow e_1 = \{-x\}$ and $\{x \geqslant 0\} = p_2 \rightarrow e_2 = \{x\}$. A more interesting example is that of computing $n!$ where n is a nonnegative integer:

$$\mathrm{fact}[\,n\,] \Leftarrow [\,n = 0 \rightarrow 1;\, n \neq 0 \rightarrow n * \mathrm{fact}[\,n-1\,]\,].$$

Note that the computation $n * \mathrm{fact}[n-1]$ must wait for the evaluation of $\mathrm{fact}[n-1]$. That is, the function of fact makes a call on itself. This type of recursion is unavailable in languages such as BASIC or FORTRAN.

To test whether a nonnegative integer n is divisible by a positive integer m we reason that

(i)   If $n = 0$, then n is divisible by m.
(ii) If $n \geqslant m$, then m divides n if and only if m divides $n - m$.
(iii) If $n < m$, then m does not divide n.

This algorithm translated into LISP becomes:

$$\mathrm{div}[\,n;m\,] \Leftarrow [\,n = 0 \rightarrow t;\, n \geqslant m \rightarrow \mathrm{div}[\,n-m;m\,];\, n < m \rightarrow f\,].$$

The letters t and f are used here as functions which have value true and false respectively.

Recursion is a powerful programming technique that the student should have at his disposal. Recursion is a tremendous aid in writing programs with just moderate complexity since such programs are much easier to read.

It is also important to note that mathematical induction can be applied to recursively defined procedures for the purpose of proving properties of programs. In this context, the procedure is referred to as recursion induction.

Programming languages such as FORTRAN and BASIC require explicit instructions to the computer. In "fact" and "div," we have been using implicit programming. That is, we are defining $n!$ in terms of $(n - 1)!$, and $\mathrm{div}[n; m]$ in terms of $\mathrm{div}[n - m; m]$. This result makes for startling clarity; it is the natural way to define many functions. Readers may wish to write similar programs in their favorite language and compare them with the above examples.

Our next example is a program that tests a polynomial defined over the integers for irreducibility. The mathematical result we use can be found in [2].

Theorem. *Assume that $q(x)$ is a polynomial, of degree d, with integer coefficients and define $S(q) = \{ \ldots, q(-1), q(0), q(1), \ldots \}$. Let p denote the number of primes and negative primes that appear in $S(q)$, and let u denote the number units (± 1's) that appear in $S(q)$. Then $q(x)$ is irreducible over the rationals if $p + 2u - d > 4$.*

The algorithm is simple enough. Generate $q(0), q(-1), q(1), q(-2), q(2), \ldots$, count and terminate when $p + 2u - d > 4$. In LISP we could write

$$\mathrm{irred}[q] \Leftarrow \mathrm{irred}'[q; 0; -\deg[q]],$$

where

$$\mathrm{irred}'[q; x; s] \Leftarrow [s > 4 \rightarrow t;$$

$$\text{"isprime"}[\mathrm{abs}[q[x]]] \rightarrow \mathrm{irred}'[q; \text{"next"}[x]; s + 1];$$
$$\text{"isunit"}[q[x]] \rightarrow \mathrm{irred}'[q; \text{"next"}[x]; s + 2];$$
$$t \rightarrow \mathrm{irred}'[q; \text{"next"}[x]; s]]$$

where "isprime" returns value true if its argument is prime, "isunit" returns value true if its argument is ± 1, and "next" generates the sequence 0, $-1, 1, -2, 2, \ldots$. We still need to write the auxiliary functions: "isprime," "next," "isunit," and "deg." For example "next" could be defined as

$$\mathrm{next}[x] \Leftarrow [x < 0 \rightarrow \mathrm{abs}[x]; x \geqslant 0 \rightarrow -(x + 1)]$$

to generate the sequence 0, $-1, 1, -2, 2, \ldots$.

The program will not terminate unless $s > 4$ at some stage of the computation. In other words, the program will terminate and we have an irreducible polynomial, or the question will remain unanswered when we have exceeded our allotted time. Unfortunately we have no practical results that give necessary and sufficient conditions for irreducibility. This is not an isolated instance of a useful program that may not terminate. Programs to prove the consistency of a set of axioms have been written (in LISP) where the program terminates precisely when the set of axioms are consistent. That is, the program terminates in a reasonable amount of time and we have our proof, or we exceed our time limit and the question is still open.

In this last example, it is necessary to be able to represent the polynomial $q(x)$ in a way that easily allows the program to evaluate $q(x)$ as well as determine its degree. Standard techniques for handling this problem lead to an interesting excursion into data structures.

The most primitive objects in LISP are numbers and strings of letters from the alphabet. These objects are called atoms. The other data type in LISP is the list (or sequence) in which the elements are either atoms or other lists. In addition to the primitive functions defined over the integers, there are five other functions which operate on lists and atoms. The function *cons* adds a single element onto the front of a list. Thus, we can build any list with repeated application of *cons* together with the set of atoms and the empty list. The function *first* selects the first element of a list, and *rest* returns all but the first element of a list. The predicate *atom* returns true if its argument is an atom, and false otherwise. Finally, *eq* is a predicate which returns true if its two arguments are the same atom, and false if the arguments are different atoms.

All other functions in LISP are defined in terms of the primitive functions and the conditional expression. For example, second[l] $\Leftarrow$ first[rest[l]], where l is a list. The simplicity of this language makes a mathematical treatment of the language straightforward and easy to learn. However, LISP is not just a toy for the theoretician; some of the most sophisticated applications of computers are written in LISP.

To actually implement our program for testing the irreducibility of a polynomial, we represent it as a list. For example the polynomial $q(x) = x^3 + 5x^2 - 7$ could be represented as $((1,3),(5,2),(-7,0))$—a list of three sublists, each of which represents a term of the polynomial. The first element of each sublist denotes the coefficient of the term and the second denotes the exponent. The degree of $q(x)$ can easily be determined: deg[q] $\Leftarrow$ maximum[second[first[q]]; deg[rest[q]]]. We also have all the information needed to evaluate $q(x)$ for varying x.

LISP has very important applications in the area of symbolic programming. One example of this is symbolic differentiation applied to polynomials of a single variable. We analyze the problem as follows:

(i)   If $p(x)$ is a constant polynomial, then $dp/dx = 0$.

(ii) If $p(x)$ is a variable, then $dp/dx = 1$.

(iii) If $p(x)$ is the sum of two polynomials $p = r + s$, then $dp/dx = dr/dx + ds/dx$.

(iv) If $p(x)$ is the product of two polynomials $p = r * s$, then $dp/dx = r * ds/dx + s * dr/dx$.

Writing the algorithm in LISP is simple:

$$\text{diff}[p; x] \Leftarrow [\text{isconst}[p] \rightarrow 0; \text{isvar}[p] \rightarrow 1;$$
$$\text{issum}[p] \rightarrow \text{sum}[\text{diff}[\text{term}_1[p]; x]; \text{diff}[\text{term}_2[p]]; x];$$
$$\text{isprod}[p] \rightarrow \text{sum}[\text{prod}[\text{term}_1[p]; \text{diff}[\text{term}_2[p]; x]];$$
$$\text{prod}[\text{term}_2[p]; \text{diff}[\text{term}_1[p]; x]]]].$$

Although the abstract functions and predicates must be written as list functions, and polynomials must be written as lists, one can always return to the abstract algorithm if complications arise. The representation problem is totally separated from the algorithm. Modifying the algorithm to include polynomials of several variables would entail changing only the second item to

$$\text{isvar}[p] \rightarrow [p = x \rightarrow 1; p \neq x \rightarrow 0].$$

The fact that we have modified our program to handle the more general case might necessitate a change in representation and a redefinition of the abstract functions. Nevertheless, the algorithm itself is only slightly altered. In other programming languages, the programmer must face both the representation problem and the algorithm at the same time. The minor changes made above might entail an entire rewrite of the program if it was heavily dependent on the representation.

Another interesting aspect of LISP is that there is no distinction between program and data. That is, the program is represented as a list and it may therefore be used as data for some other program. This property of LISP provides a good amount of insight into computer science. For example, one can write an interpreter for LISP without the need to learn another language or do an in-depth study of a particular machine. One needs to understand the evaluation process, and then write an algorithm in LISP which does the evaluation and write any abstract functions in terms of LISP functions. Other language programs constitute the data for this program; the representation problem has already been taken care of by the LISP system.

The above exercise provides motivation for a discussion of storage structures (such as linked lists, stacks, and symbol tables) needed for its implementation. Topics such as garbage collection, scanning and parsing may also be discussed by writing such algorithms in LISP. Students who finish this study will not only understand LISP; they will have a deeper appreciation of programming languages in general.

One of the most important differences between LISP and other programming languages is the fact that it is a mathematical formalism as well. LISP has a mathematical foundation for proving properties of programs, equivalence of programs, and correctness of programs. This provides faculty with motivation and material for developing courses in logic and computer applications such as theorem proving and proof checking.

This material has been taught at San Jose State as a one-semester graduate course for the Master's Degree program in Computer and Information Science. It would also be extremely valuable as a one-year course for undergraduate math majors having had only an introductory programming course and perhaps some calculus background. A complete discussion of these topics may be found in [1]. This source would also serve as an excellent text for such a course.

REFERENCES

1. John Allen, The Anatomy of LISP, to be published by McGraw-Hill, New York, 1979.
2. W. S. Brown and R. L. Graham, An irreducibility criterion for polynomials over the integers, Amer. Math. Monthly, 76 (Aug.-Sept. 1969) 795–797.
3. Thomas E. Cheathan, Jr., The unexpected impact of computers on science and mathematics, Proc. Symposia Appl. Math., 20 (1974) 133–137.
4. E. W. Dijkstra, The humble programmer, Comm. ACM, Vol. 15, No. 10 (Oct. 1972) 859–866.
5. John McCarthy, A Basis for a Mathematical Theory of Computation, Computer Programming and Informal Systems, North Holland, Amsterdam, 1963, pp. 33–69.
6. Paul S. Wang, A symbolic manipulation system, Proc. Internat. Comput. Symposium, 1 McGraw-Hill, New York, 1978, pp. 103–109.

Simulating Sampling from Normal Populations

Catherine Lilly, *Westfield State College, Westfield, MA*

In statistics classes, one often speaks of obtaining a random sample of values from a normal population. If there is not sufficient time or resources to actually collect large samples (of weights, heights, measurements, etc.), it is difficult for the student to visualize exactly what such a sample would look like. Usually, information about the distribution of values in a normal population is conveyed by a sketch of the normal curve (Figure A).

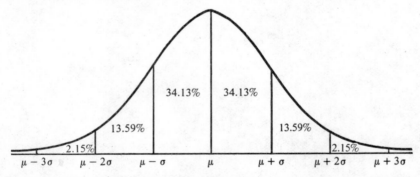

Percentages of measurements falling within an interval that is one standard deviation wide.

Figure A.

However, this is insufficient to illustrate certain facts about samples from normal populations, facts which can best be visualized from direct examination of numerical samples. For example, if IQ's are normally distributed with mean 100 and standard deviation 15, and if 100 random persons are asked their IQ's, how many will be exactly 100? How many will be above 150? Figure B illustrates what one set of answers to these questions might be.

If one hundred random z-scores are computed from a normal population, how regular is the pattern of alternation between positive and negative values? Figure C illustrates the pattern that might occur in a sample of normally distributed values.

	$\mu = 100$	$\sigma = 15$	$n = 100$	
97	101	78	88	91
97	110	90	108	93
114	107	93	89	83
74	73	102	93	107
93	123	130	104	91
87	80	91	97	108
72	93	118	89	77
98	82	93	81	116
80	88	108	113	118
121	103	108	96	103
89	106	95	122	88
88	101	127	83	85
106	91	134	82	97
125	99	86	91	130
96	93	75	77	105
109	93	113	82	70
99	109	120	128	113
88	109	83	96	102
93	97	125	117	84
102	108	93	98	109

Numerical sample of 100 values simulating a normal distribution with mean of 100 and standard deviation of 15.

Figure B.

	$\mu = 0$	$\sigma = 1$	$n = 100$	
− 1.6	1.21	−.13	−.68	.42
.82	−.44	−.46	.39	−.08
3.32	1.61	.14	−.45	− 1.48
.3	−.24	−.13	−.42	.83
−.34	1.81	−.43	1.93	.18
−.43	−.23	− 1.02	−.32	−.68
− 1.74	2.05	−.34	1.71	.07
.54	.79	.02	.19	−.25
−.44	−.77	−.74	−.03	− 1.13
.58	−.92	.09	−.41	−.25
−.25	.18	.18	.37	.23
.59	−.22	− 1.35	.48	−.46
− 1.7	−.03	.02	.74	−.98
−.01	1.03	.92	− 1.83	1.21
1.44	−.01	1.03	−.93	1.04
1.11	1.55	−.91	−.43	.27
1.39	.4	.78	−.1	.06
.87	.57	.97	− 1.51	−.8
2.06	−.45	−.35	0	− .3
.94	−.9	.41	−.16	.76

Numerical sample of 100 values simulating a normal distribution with mean of 0 and standard deviation of 1.

Figure C.

Since students benefit from learning to handle large masses of data, it is useful for instructors to be able to generate samples such as those above, for analysis by students. However, standard reference works do not contain sample tables of this type, and since physical collection of data is often too time-consuming, a valuable alternative, which was used in the two examples above, is to be able to simulate the sampling process and generate tables of this type on the computer. This can be done on any computer with a random number generator.

Suppose that one wishes to construct a table consisting of numbers selected randomly from a normal population with mean μ and standard deviation σ. Assume further that one wants these numbers to come from the interval $(\mu - t\sigma, \mu + t\sigma)$, where t is chosen sufficiently large so as to insure that the probability of a number occurring outside this interval is essentially nil. The set of random numbers selected from this interval has

$$\text{Mean } \mu_r = \mu$$

$$\text{Variance } \sigma_r^2 = \frac{1}{2\sigma t} \int_{\mu - \sigma t}^{\mu + \sigma t} (x - \mu)^2 \, dx = \frac{\sigma^2 t^2}{3}$$

$$\text{Standard deviation } \sigma_r = \frac{\sigma t}{\sqrt{3}}.$$

Suppose, for example, that samples of size n ($n > 30$) are chosen from this population. By the Central Limit Theorem, the means of these samples will be approximately normally distributed with

$$\text{Mean } \mu_m = \mu$$

$$\text{Standard Deviation } \sigma_m = \frac{\sigma_r}{\sqrt{n}} = \frac{\sigma t}{\sqrt{3n}}.$$

Thus, if one takes $n > 30$ and chooses sufficiently large t so that $\sigma_m = \sigma t / \sqrt{3n} = \sigma$ (i.e., $t^2/3 = n$), then the distribution of means of samples of size n chosen from a random population in the range $(\mu - \sigma t, \mu + \sigma t)$ will be approximately normal with mean and standard deviation equalling the desired μ and σ.

The following computer program written in BASIC simulates the process of sampling from a normal population, based on the ideas outlined above. The values used here are $n = 48$ and $t = 12$. Line 146 or 145, but not both, should be used, depending on whether one desires real or integer-valued output.

```
5 REMARK SIMULATING SAMPLING FROM NORMAL POPULATIONS
10 PRINT "WHAT MEAN AND STANDARD DEVIATION";
20 INPUT M, S
30 PRINT "HOW MANY VALUES ARE DESIRED";
40 INPUT Q
50 T = 12
60 N = 48
```

```
70 PRINT "MEAN ="; M, "STANDARD DEVIATION ="; S, Q; "VALUES"
80 FOR J = 1 TO Q
90 A = 0
95 REMARK GENERATE N RANDOM VALUES AND OUTPUT THEIR MEAN
100 FOR I = 1 TO N
105 REMARK P IS A RANDOM VALUE ON THE INTERVAL (M - TS, M + TS)
110 P = 2 * S * T * RND(-1) + (M - S * T)
120 A = A + P
130 NEXT I
135 REMARK USE LINE 145 FOR INTEGER OUTPUT
136 REMARK USE LINE 146 FOR REAL OUTPUT ROUNDED TO TWO DECIMAL
PLACES
137 REMARK DO NOT USE BOTH LINES
145 PRINT INT(A/N + .5)
146 PRINT INT(A/N * 100)/100
150 NEXT J
160 END
```

Extended Precision Arithmetic— A Numeric Approach

Gerard Kiernan, *Manhattanville College, Purchase, NY*

Factorial N, usually written $N!$, appears in a variety of ways in introductory courses. Binomial coefficients, permutations and combinations all involve factorials. The computation of $N!$, even for relatively small values of N, involves extremely large numbers. For example, the number of different ways to arrange a standard deck of playing cards is 52! and this is a 68-digit number. Any attempt to compute this number exactly shows how difficult it is, even in this age of the computer, to work with extremely large numbers.

Every computer has a built-in precision for arithmetic computations. Most computers approximate large or small numbers using exponential notation, and they have a relatively small range of exact values. Thus, problems that require a greater degree of precision need some techniques to extend the built-in precision limitations.

Special techniques have been devised to deal with specific problems [3]. The purpose of this article is to examine a general procedure for obtaining accurate results when working with large numbers.

The basic "pencil and paper" method of multiplication provides an algorithm for finding the product of two numbers with any number of digits. A slight variation of this algorithm can be used to write a computer program to find the product of large numbers.

The "pencil and paper" method of multiplication simply requires that we can find:

1. the product of two one-digit numbers,
2. the sum of two one-digit numbers.

If we use the fact that the computer can find:

1. the product of two N-digit numbers,
2. the sum of two N-digit numbers,

then the "pencil and paper" algorithm can be improved by using a base of 10^N rather than a base of 10 (see [1]).

If the base is a power of 10, there are no conversion problems. In fact, the number in base 10^N can be read directly from the number in base 10 by taking the base 10 digits in groups of N. The number 11223344556677 is a 14-digit number in base 10. In base 10^6, it becomes a 3-digit number with digits 11,223344 and 556677.

The "pencil and paper" method of multiplication works the same way in base 10^N as it does in base 10. By using base 10^N (rather than base 10), the number of operations required in the "pencil and paper" algorithm can be reduced. For example, the number of multiplications can be reduced by a factor of N^2. In a computer where the product of two numbers is accurate to D digits, let $N = [D/2]$ (the greatest-integer less than or equal to $D/2$).

The program MULT is a general program (written in MICROS extensive BASIC) for the product of two integers that uses the "pencil and paper" algorithm in base 10^N. By using the extended precision mode of this system (this is the LONG statement in line 140 of the program), the product of two integers is accurate to 12 digits. So $N = [12/2] = 6$. The base is set at 10^6 in line 20.

The listing of the program MULT gives an example of how this program is used to find the product of two large integers. In base 10, the first number

$$U = 7234367859408761245670764512553679$$

has 34 digits; the second number

$$V = 63103347849907632241356763511$$

has 29 digits. In base 10^6, however, U is a 6-digit number and V is a 5-digit number. The numbers of digits in U and V, in base 10^6, are entered in lines 50 and 60 as DATA statements.

The "pencil and paper" methods for addition, subtraction and division also provide algorithms for working with large numbers. As in the case of multiplication, these algorithms work equally well in base 10^N (see [2]).

In most cases, programs based on these algorithms would be used as subroutines in larger programs. An example of this is given in the program FACT, which computes $N!$. This program uses the method of the program MULT, as a subroutine, in lines 130 to 210. The run of this program with $N = 52$ gives the exact value of 52!

```
0001 REM PROGRAM MULT
0010 PRINT
0020 LET B=10^6
0030 LET N=6
0040 LET M=5
0050 DATA 7234,367859,408761,245670,764512,553679
0060 DATA 63103,347849,907632,241356,763511
0070 DIM U(N),V(M)
0080 FOR I=1 TO N
0090   READ U(I)
0100 NEXT I
0110 FOR I=1 TO M
```

```
0120   READ V(I)
0130 NEXT I
0140 LONG W(N+M),T,K
0150 FOR I=M+1 TO M+N
0160   LET W(I)=0
0170 NEXT I
0180 FOR J=M TO 1 STEP -1
0190   IF V(J)><0 THEN GOTO 0220
0200   LET W(J)=0
0210   GOTO 0290
0220   LET K=0
0230   FOR I=N TO 1 STEP -1
0240     LET T=U(I)*V(J)+W(I+J)+K
0250     LET W(I+J)=T-(INT(T/B)*B)
0260     LET K=INT(T/B)
0270   NEXT I
0280   LET W(J)=K
0290 NEXT J
0300 PRINT
0310 PRINT "U*V="
0320 LET C,Z=0
0330 LET C=C+1
0340 IF W(C)=0 THEN GOTO 0330
0350 PRINT W(C);
0360 FOR I=C+1 TO N+M
0370   PRINT USING "999999",W(I);
0380   LET Z=Z+.2
0390   IF Z<1 THEN GOTO 0420
0400   PRINT
0410   LET Z=0
0420 NEXT I
0430 END


RUN

U*V=
 456 512831506462733878787303747270064810
565074801117493096006969



0001 REM PROGRAM FACT
0010 LONG T,K
0020 PRINT
0030 PRINT "INPUT N FOR N!"
0040 INPUT R
0050 LET L=INT(LOG(((R/2.71828)^R)*(6.28318*R))/LOG(10))
0060 LET L=INT(L/6)+1
0070 LET B=10^6
0080 LET A,M,N=1
0090 DIM U(L+1),W(L+1)
0100 FOR V=1 TO R
0110   IF V>1 THEN GOTO 0130
0120   LET U(V)=1
```

```
0130    FOR I=2 TO N+1
0140       LET W(I)=0
0150    NEXT I
0160    LET K=0
0170    FOR I=N TO 1 STEP -1
0180       LET T=U(I)*V+W(I+1)+K
0190       LET W(I+1)=T-(INT(T/B)*B)
0200       LET K=INT(T/B)
0210    NEXT I
0220    LET W(1)=K
0230    LET Z=2
0240    IF W(1)=0 THEN GOTO 0270
0250    LET Z=Z-1
0260    LET N=N+1
0270    FOR I=Z TO N+1
0280       LET U(I-Z+1)=W(I)
0290    NEXT I
0300 NEXT V
0305 PRINT
0310 PRINT R;"!="
0320 PRINT
0330 LET C,Z=0
0340 LET C=C+1
0350 IF W(C)=0 THEN GOTO 0340
0360 PRINT W(C);
0370 FOR I=C+1 TO N+1
0380    PRINT USING "999999",W(I),
0390    LET Z=Z+.2
0400    IF Z<1 THEN GOTO 0430
0410    PRINT
0420    LET Z=0
0430 NEXT I
0440 PRINT
0450 PRINT
0460 END


RUN
INPUT N FOR N!
?52
  52 !=

  80  65817517094387857166063685640376975
2895054408832778240000000000000
```

REFERENCES

1. J. Ballantine, Note on the multiplication of long decimals, Amer. Math. Monthly, 30 (1923) 68.
2. D. Knuth, The Art of Computer Programming: Seminumerical Algorithms, Vol. 2, Addison-Wesley, Reading, Mass., 1968.
3. P. Scopellitti and H. Peebles, Finding super accurate integers, The Two-Year College Mathematics Journal, 7 (Sept. 1976) 52–54.

7. MATHEMATICS EDUCATION

> *"There are things which seem incredible to most men who have not studied mathematics."*
>
> —Archimedes

> *"Some persons have contended that mathematics ought to be taught by making illustrations obvious to the senses. Nothing can be more absurd or injurious: it ought to be our never-ceasing effort to make people think, not feel."*
>
> —S. T. Coleridge, Lectures on Shakespeare (Bohn Library), 52

> *"The number of mathematical students ... would be much augmented if those who hold the highest rank in science would condescend to give more effective assistance in clearing the elements of the difficulties which they present."*
>
> —A. DeMorgan, Study and Difficulties of Mathematics (Chicago, 1902), Preface

How does one define good teaching? Is it the ability to explain abstract concepts in more familiar experiential terms? Does it mean skill in revealing patterns whose conjectured formulations can be readily corroborated? Is it the successful training of students to become effective problem solvers? The answer, of course, is that good teaching encompasses all of these characteristics, and much more.

There are probably three primary components of teaching. The first is to *instruct*—to impart knowledge in a methodical and formal manner. The second is to *educate*—that is, to bring out latent capabilities which enable students to come to know the subject. And the third is to *train*—meaning to instruct and drill with specific ends in view. The manner in which these components are brought to bear on teaching can, and does, vary considerably depending on who is being taught what, and for which purpose. Since the study of mathematics involves different types of objects, systems, and reasoning processes, and since students vary considerably in the requisite cognitive skills, it should be clear that the best way to teach one topic may not be the best way to teach a different topic.

On one point we can all agree: A good math teacher, like a talented chef, can stimulate students' curiosity, whet their intellectual appetites and create a smorgasbord of interesting food for thought. In this context, teachers can use the following articles to spice up their presentations and add flavor to the classroom learning experience.

One effective teaching technique in mathematics is to relate a given symbolic system to another which is more concrete and less abstract. "A Logical Approach to Teaching Formal Logic," by G. Vervoort, is a good case in point. His article describes a classroom lesson that visually relates formal logic to multiplication modulo 8. The author shows at which point the notion of isomorphism can be introduced and matters such as associativity and commutativity of the original system can be considered.

Recognizing and verifying patterns is an important facet of mathematical endeavor. The next two articles illustrate this process, of analysis and proof, by which mathematics is created. In "The Search for Sums," S. Avital and E. Theil demonstrate how numerical patterns, naturally associated with sums of series, lead to algebraic conjectures which can usually be proved by mathematical induction. The next article, by James Friel, gives balance and adds dimensions to this discussion. In "An Elementary Proof for a Famous Counting Problem," an interesting geometric problem quickly leads students to a pattern whose general formulation turns out to be wrong. Their investment in this "nonpayoff" serves as the stimulus for students to discover what the true pattern really should be.

How can we convey the essence—the dynamics and the fascination—of mathematics to our students? One way is through the challenge and excitement of doing mathematical research—interpreted here to mean investigating (by close examination or experimentation) the nature of, or solution to, a problem of interest. Fruitful research exists at all levels. John Staib demonstrates this in "Problem-Solving versus Answer-Finding." Here the class (and instructor) adopt a research attitude, and thus discover ten different precalculus solutions to a standard textbook problem. Next, Carolyn Shevokas Funk describes how students can express a problem as an experiment in probability, and then repeatedly perform this experiment, with the aid of a computer, in order to estimate the solution to the original problem. Her article, "Using Monte Carlo Methods to Solve Probability Problems," studies the success and satisfaction of students in using their ingenuity to solve complicated problems whose answers were not known.

There are numerous articles (of which the preceding is one) that illustrate the computer's use in analysis courses. Rarely, however, is the computer used in broad-based Liberal Arts courses. Thus, the time is ripe for Gordon, Meyer, and Shindhelm's article "Introducing the Computer in Liberal Arts Mathematics." Here the authors focus on how computers can be effectively used, in survey of mathematics courses, to enhance topics taught (number theory, group theory, symbolic logic) and to deepen students' appreciation of the concepts being developed.

Everything considered thus far has, of course, been based on the assumption that students have the prerequisite knowledge and ability to understand the concepts being studied. Unfortunately, this is not always the case. Standardized placement tests frequently do not meet the needs of individual schools, and many who design their own tests are not always sure that the constructed tests are reliable measures of present knowledge or reliable predictors of future success. Since erroneous placement impedes learning and teaching, research aimed at improving such predictive tests are of value to the whole educational process. In "Mathematics Placement Test Construction," F. Pedersen discusses some practical aspects of test design that were formulated while he was constructing a placement and proficiency test at Southern Illinois University. Pedersen's process of test construction can be used by mathematicians, who may not have extensive training in computers and test construction design, to develop placement tests tailored to the specific needs of their campus.

A Logical Approach to Teaching Formal Logic

G. Vervoort, *Lakehead University, Thunder Bay, Ontario, Canada*

Students frequently feel that the mathematics of formal logic is dull and useless—especially if their text presents this topic in a separate chapter and never refers to it again. The following classroom lesson, relating logic to multiplication modulo 8, usually captivates students' interest and helps to create a better impression of the value and use of formal logic.

While students are still wondering whether it is really worth learning formal logic, ask them to complete the table below.

*	$(p \rightarrow q)$ statement	$(q \rightarrow p)$ converse	$(\sim p \rightarrow \sim q)$ inverse	$(\sim q \rightarrow \sim p)$ contrapositive
statement				
converse				
inverse				converse
contrapositive				

This table is read from left to right. Thus to find the inverse of the contrapositive, students must negate both the hypothesis and the conclusion of the contrapositive. The result is the converse.

Students may groan while filling in the blanks, but they will endure it if they are assured that the table will be put to good use in a little while. The very drudgery of the task is part of the intended learning experience since it enhances appreciation of what follows later. Allow 10–15 minutes for this; they need time to practice and become familiar with these relatively new concepts. (Some will observe that each conditional occurs exactly once in each row and in each column.) Display the completed results, using abbreviations (Table 1), for everyone.

Next, have the class complete the multiplication table modulo 8 on the domain $\{1, 3, 5, 7\}$. Students will complete this exercise quickly if they are reminded that multiplication modulo 8 means "Divide the product by 8

206

Table 1.

*	st	cv	iv	cp
st	st	cv	iv	cp
cv	cv	st	cp	iv
iv	iv	cp	st	cv
cp	cp	iv	cv	st

and write the remainder." In contrast to the previous task, this one will seem trivial. Display the completed results (Table 2) for everyone.

Table 2.

⊗	1	3	5	7
1	1	3	5	7
3	3	1	7	5
5	5	7	1	3
7	7	5	3	1

After as many students as possible have been drawn into commenting on how much easier it was to complete Table 2 than Table 1, superimpose the two tables on top of each other, and invite the class to discover the relationship between the numbers and the conditionals. The one-to-one correspondence in Table 3 will leap right at them.

Table 3.

*	⊗	st	1	cv	3	iv	5	cp	7
st	1	st	1	cv	3	iv	5	cp	7
cv	3	cv	3	st	1	cp	7	iv	5
iv	5	iv	5	cp	7	st	1	cv	3
cp	7	cp	7	iv	5	cv	3	st	1

Now students can see that to solve a problem such as finding the converse of the inverse of the contrapositive (in case anyone of sound mind ever was insane enough to want to know), one need only go through a translation process:

cv	*	(iv	*	cp)	=	?
↓	↓	↓	↓	↓	↓	↑
3	⊗	(5	⊗	7)	=	1

Applying these substitution procedures to a few complicated statements helps the student to quickly see the value of such a process (and provides a rationale for practicing the conditionals again).

The matter of associativity and commutativity of the original system can now be settled since students can see that multiplication modulo 8 has these properties. In better classes, the notion of isomorphism may be introduced. When do we make use of isomorphisms? Conceivably if one had to design a computer to deal with conditionals, one might wire in such translations. Another useful application of this concept of isomorphism was discovered by Napier when he substituted simple addition for cumbersome multliplication and invented logarithms over $\mathbb{R}^+ = (0, \infty)$:

$$\log : (\mathbb{R}^+, \cdot) \to (\mathbb{R}^+, +)$$
$$x \to \log x$$
$$y \to \log y$$

where

$$\log(x \cdot y) = \log x + \log y.$$

By now the bell should be about to ring. Assign students the task of finding other one-to-one correspondences between the conditionals and multiplication modulo 8 that work equally well. Real enthusiasts can look for an isomorphism between the conditionals and a system of four movements $\{I, R_{180}, D_1, D_2\}$ of a square. Don't tell them! Class dismissed.

The Search for Sums

S. Avital, *Israel Institute of Technology, Technion, Haifa, Israel,* and
Edward Theil, *University of California, Berkeley, CA*

Students encounter general sequences and series when dealing with applications of mathematical induction. In that topic, the outcomes and results are usually presented and the student is then asked to "prove" that for any integral $n > 0$,

$$1^2 + 2^2 + 3^2 + \cdots + n^2 = n(n + 1)(2n + 1)/6$$

or to "show" that, for any positive integer n,

$$2^3 + 4^3 + 6^3 + \cdots + (2n)^3 = 2n^2(n + 1)^2.$$

It is important to recognize that mathematical induction is here an accessory after the fact. Someone originally conjectured that the required sum can be obtained from a certain formula, and the student is only required to prove it. One may well ask, *"For heaven's sake, how do you guess the conjecture in the first place?"*

Mathematics is part of human culture and, as such, its cultural values lie not so much in its applications as in the methodology and the flexibility of its approach. To give the student some feeling for this, we have to expose him to the ideas that are used to generate theorems. Moreover, when an idea is used, we should try to show that the same approach is applicable in more than one situation. In contrast, when proving the formula for the sum of the geometric series, most textbooks use an approach which is neither repeated nor used to find the sum of another series. This means that the approach to the proof is generally forgotten, and a major objective for teaching such proofs is not attained.

The search for sums is, like much of mathematics, a search for patterns. One first looks for a pattern in numerical examples. When a pattern becomes evident, one formulates a conjecture and tests this with additional numerical examples. If the results support the conjecture, the final proof (usually by mathematical induction) can then be attempted. This is frequently the process by which mathematics is created.

After each example, we have inserted additional problems that are closely related to the example being discussed, and which can be explored by the reader in an inductive numerical manner before an algebraic conjecture is made. These problems are labelled E1, E2, E3, and so on.

I. Using ratios. We shall begin with problems whose solutions are well known.

1. Find the sum of the series $S_n = 1 + 2 + 3 + \cdots + n$.

An inductive investigation produces:

n	1	2	3	4	5
S_n	1	$1 + 2 = 3$	$1 + 2 + 3 = 6$	$1 + 2 + 3 + 4 = 10$	$1 + 2 + 3 + 4 + 5 = 15$

It stands to reason that the sum S_n is a function of n, and so we shall look for a simple relationship between S_n and n. It turns out that a multiplicative relationship produces promising results.

Looking at the ratio S_n/n,

n	1	2	3	4	5
S_N/n	1/1	3/2	$6/3 = 2$	$10/4 = 5/2$	$15/5 = 3$

and writing 1 as 2/2, and 2 as 4/2, and 3 as 6/2, we notice a pattern which leads to the conjecture that $S_n/n = (n + 1)/2$. A further check for $n = 6$ produces $S_6/6 = 21/6 = 7/2$. Similarly, $n = 7$ yields $S_7/7 = 28/7 = 4 = 8/2$. The results support the conjecture:

$$S_n/n = (n + 1)/2 \quad \text{or} \quad S_n = n(n + 1)/2.$$

A proof, by mathematical induction, or other means, can now be attempted. (The reader may know that the numbers $1, 3, 6, 10, \ldots$ generated by $n(n + 1)/2$ are called triangular numbers.)

E1: It might be interesting, in a classroom setting, to similarly investigate the sums $S_n^{(\text{odd})} = 1 + 3 + 5 + \cdots + (2n - 1)$ and $S_n^{(\text{even})} = 2 + 4 + 6 + \cdots + 2n$.

2. Let us try a similar approach for the series $1^2 + 2^2 + 3^2 + \cdots + n^2 = S_n^{(2)}$. Again, an inductive investigation produces the following results:

n	1	2	3	4	5
$S_n^{(2)}$	1	5	14	30	55

Since an investigation of $S_n^{(2)}/n$ does not produce a recognizable pattern, we try the ratio $S_n^{(2)}/S_n$.*

n	1	2	3	4	5
S_n	1	3	6	10	15
$S_n^{(2)}$	1	5	14	30	55
$S_n^{(2)}/S_n$	1/1	5/3	14/6 = 7/3	30/10 = 3	55/15 = 11/3

Observing that $1/1 = 3/3$ and $3 = 9/3$, we notice a pattern. This should lead to the hypothesis that $S_n^{(2)}/S_n = (2n + 1)/3$. A check of two more results produces $S_n^{(2)}/S_6 = 91/21 = 13/3$ and $S_7^{(2)}/S_7 = 140/28 = 5 = 15/3$. Encouraged by these results, we conjecture that: $S_n^{(2)}/S_n = (2n + 1)/3$. Hence

$$S_n^{(2)} = S_n(2n + 1)/3 = n(n + 1)(2n + 1)/6.$$

We can now attempt to produce the general proof by mathematical induction.

E2: The reader may similarly investigate the sum of squares of odd natural numbers and of even natural numbers. Note that either one of these sums can be obtained from the other by simple subtraction from $S_n^{(2)}$.

3. It is natural now to investigate the sum $S_n^{(3)}$ of the cubes of the natural numbers. (In this example the pattern of $S_n^{(3)}$ is also obviously a sequence of certain squares.) Following the previous result, it stands to reason that we look immediately at the ratios $S_n^{(3)}/S_n^{(2)}$ or $S_n^{(3)}/S_n$.

n	1	2	3	4	5
S_n	1	3	6	10	15
$S_n^{(2)}$	1	5	14	30	55
$S_n^{(3)}$	1	9	36	100	225
$S_n^{(3)}/S_n$	1	3	6	10	15
$S_n^{(3)}/S_n^{(2)}$	1/1	9/5	36/14 = 18/7	100/30 = 10/3	225/55 = 45/11

---

*The idea of considering the ratio $S_n^{(2)}/S_n$ was used by G. Pólya in *Induction and Analogy in Mathematics*, Princeton University Press, 1954, p. 108.

We immediately see that $S_n^{(3)}/S_n$ seems to fit the pattern of triangular numbers. In other words, it appears that

$$S_n^{(3)} = S_n\left[\frac{n(n+1)}{2}\right] = \left[\frac{n(n+1)}{2}\right]^2.$$

After a little experimenting with the ratio $S_n^{(3)}/S_n^{(2)}$, we write $1/1$ as $3/3$, $10/3$ as $30/9$, and the pattern in the denominator is seen to be 3, 5, 7, ..., $(2n+1)$. The pattern in the numerator is 3, 9, 18, 30, 45 or 3×1, 3×3, 3×6, 3×10, 3×15—that is, three times a corresponding triangular number. It seems that

$$S_n^{(3)}/S_n^{(2)} = \left[\frac{3n(n+1)}{2}\right]/(2n+1).$$

Checking this for $n = 5$ and $n = 7$, we see that

$$S_6^{(3)}/S_6^{(2)} = 441/91 = 63/13 = 3 \times 21/13 \quad \text{and} \quad S_7^{(2)} = 140,$$

and

$$S_7^{(3)}/S_7^{(2)} = 784/140 = 28/5 = 3 \times 28/15$$

fits the pattern. Therefore we conjecture that

$$S_n^{(3)}/S_n^{(2)} = \left[\frac{3n(n+1)}{2}\right]/(2n+1),$$

or

$$S_n^{(3)} = \frac{n(n+1)(2n+1)}{6} \cdot \frac{3n(n+1)}{2n(2n+1)} = \frac{n^2(n+1)^2}{4} = \left[\frac{n(n+1)}{2}\right]^2.$$

We can now attempt to prove this result by mathematical induction.

E3:   The reader can similarly obtain formulas for the sum of the cubes of n even integers, or n odd integers.

4.   Does this ratio approach work for other series? To formulate a conjecture for the sum of n terms of the series

$$W_n = (1 \times 2) + (2 \times 3) + (3 \times 4) + \cdots + n(n+1), *$$

We shall consider the ratio between W_n and S_n.

---

*Notice that the sum can easily be obtained from $\sum_1^n k(k+1) = \sum_1^n (k^2 + k)$. Nevertheless, we think it important to show that the result can be obtained through the ratio approach—as the conjecture can be investigated in lower grades, where the more formal derivation would be difficult.

n	1	2	3	4	5	6
W_n	2	8	20	40	70	112
S_n	1	3	6	10	15	21
W_n/S_n	2	8/3	10/3	4	14/3	16/3

Representing 2 as $6/3$ and 4 as $12/3$, it appears that $W_n/S_n = (2n + 4)/3$. Since $W_7/S_7 = 168/28 = 6 = 18/3$ and $W_8/S_8 = 240/36 = 20/3$ fits this pattern, we hypothesize that $W_n/S_n = (2n + 4)/3$ or $W_n = S_n(2n + 4)/3 = n(n + 1)(n + 2)/3$.

Again, the final proof may be obtained by induction.

E4: We suggest that the reader investigates in the same way:

$$1 \cdot 3 + 2 \cdot 5 + 3 \cdot 7 + \cdots + n \cdot (2n + 1),$$
$$1 \cdot 2 + 2 \cdot 4 + 3 \cdot 6 + \cdots + n \cdot (2n) \quad \text{or}$$
$$1 \cdot 2 \cdot 3 + 2 \cdot 3 \cdot 4 + 3 \cdot 4 \cdot 5 + \cdots + n \cdot (n + 1) \cdot (n + 2).$$

Notice that some of these (as for instance the one marked (*)) can be obtained in a much shorter way. The reader should also generate some new series and check whether he can formulate a conjecture about the sum of n terms.

II. Using subtractions.

In the next few examples of series summation, we arrive at a conjecture in a way similar to the approach used in most textbooks for the summation of the geometric series.

5. Find the sum of

$$T_n = 1 + 2(\tfrac{1}{2}) + 3(\tfrac{1}{2})^2 + 4\left(\frac{1}{2}\right)^3 + \cdots + n(\tfrac{1}{2})^{n-1}.$$

It is easy to see that if we multiply each term by $1/2$, we will obtain a similar expression. Hopefully, doing this and subtracting one series from the other yields a simpler sum. From

$$T_n = 1 + 2(\tfrac{1}{2}) + 3(\tfrac{1}{2})^2 + 4(\tfrac{1}{2})^3 + \cdots + n(\tfrac{1}{2})^{n-1},$$

subtract

$$\tfrac{1}{2}T_n = 1(\tfrac{1}{2}) + 2(\tfrac{1}{2})^2 + 3(\tfrac{1}{2})^3 + \cdots + (n - 1)(\tfrac{1}{2})^{n-1} + n(\tfrac{1}{2})^n,$$

to obtain

$$\tfrac{1}{2}T_n = 1 + (\tfrac{1}{2}) + (\tfrac{1}{2})^2 + (\tfrac{1}{2})^3 + \cdots + (\tfrac{1}{2})^{n-1} - n(\tfrac{1}{2})^n.$$

The series on the right-hand side of the last equation contains (except for the last term) the geometric series with ratio $1/2$. Thus, we may write

$$\tfrac{1}{2}T_n = \frac{1 - (\tfrac{1}{2})^n}{1 - \tfrac{1}{2}} - n(\tfrac{1}{2})^n$$

and

$$T_n = 4\left[1 - (\tfrac{1}{2})^n\right] - \frac{n}{2^{n-1}} = \frac{2^{n+1} - 2 - n}{2^{n-1}}. \tag{1}$$

It is interesting to note that (1) can be derived by the ratio method used in the previous section. To do this, compare T_n with the corresponding geometric series $G_n = 1 + (1/2) + (1/2^2) + \cdots + (1/2^{n-1})$:

n	1	2	3	4	5
G_n	1	3/2	7/4	15/8	31/16
T_n	1	2	11/4	13/4	57/16
T_n/G_n	1/1	4/3	11/7	26/15	57/31

Some experimentation shows that, in each case, the numerator is twice the denominator less a simply determined integer. Thus,

$$n = 1: \quad 1 = 2(1) - 1,$$
$$n = 2: \quad 4 = 2(3) - 2,$$
$$n = 3: \quad 11 = 2(7) - 3,$$
$$n = 4: \quad 26 = 2(15) - 4.$$

We therefore conjecture that

$$\frac{T_n}{G_n} = \frac{2(2^n - 1) - n}{2^n - 1}.$$

We leave the rest to the reader.

E5: The above procedure can be extended. Multiplication by k, and subtraction, will yield a conjecture about the sum of

$$1 + 2k + 3k^2 + 4k^3 + \cdots + nk^{n-1} \tag{1a}$$

for an arbitrary k.

E6: The reader can explore the series:

$$T_n^{(2)} = 1^2 + 2^2(\tfrac{1}{2}) + 3^2(\tfrac{1}{2})^2 + 4^2(\tfrac{1}{2})^3 + 5^2(\tfrac{1}{2})^4 + \cdots + n^2(\tfrac{1}{2})^{n-1}.$$

Use the subtraction method to obtain a simpler series which can easily be summed. (Note: The series obtained in this way will be a combination of T_n and G_n.)

E7: Proceeding in the same direction, the subtraction method can be used to conjecture the sum of

$$T_n^{(3)} = 1^3 + 2^3(\tfrac{1}{2}) + 3^3(\tfrac{1}{2})^2 + 4^3(\tfrac{1}{2})^3 + \cdots + n^3(\tfrac{1}{2})^{n-1}.$$

(This time, the "simpler" series will be a combination of $T_n^{(2)}$, T_n and G_n.)

Thus, again, one can see a pattern developing in this family of series, and the interested reader may be able to develop his or her own generalizations.

Another path to follow from formula (1) is to consider the infinite series obtained from T_n as $n \to \infty$ (that is, as n increases without bound). In calculus texts, the emphasis is always on tests for convergence or divergence of infinite series, but a direct examination of (1) leads to the conjecture that the sum of the infinite series is

$$1 + 2(\tfrac{1}{2}) + 3(\tfrac{1}{2})^2 + 4(\tfrac{1}{2})^3 + 5(\tfrac{1}{2})^4 + \cdots = 4.$$

In fact, the subtraction method is actually easier to use on an infinite series than on its finite counterpart, because there is no last term in the former. (Caution: the method is invalid if the infinite series is divergent!)

III. An unusual series. So far, our attention has been directed toward series whose nth term a_n can be found easily by informal deduction. For example, the series $S_n^{(2)}$ had $a_n = n^2$, and the series T_n had $a_n = n(1/2)^{n-1}$. However, one can easily write a series whose terms are quite obvious, and yet find it difficult to obtain an explicit formula for a_n. For example, consider $1 + 2 + 2 + 3 + 3 + 3 + 4 + 4 + 4 + 4 + 5 + \cdots$. Although it is clear that there will be five 5's, six 6's and so on, how are we to determine a_n (other than by counting)? Once again, our method is one of inductive exploration.

Because each integer appears only as part of a string of consecutive equal integers, we start by making a table which tells us where each string starts and where it ends.

When the nth term equals	the string starts at position	the string ends at position
1	1	1
2	2	3
3	4	6
4	7	10
5	11	15
6	16	21
7	22	28

Let us denote the nth term by a_n, each starting position as n_s, and each ending position as n_e. Can we find a rule to determine the values of n_s and n_e?

We see that the n_e are just the triangular numbers, the sums of our very first series S_n. In other words, we conjecture that the last appearance of the number a_n occurs at position

$$n_e = \frac{a_n(a_n + 1)}{2} \tag{2}$$

in the series.

The sequence for n_s is 1, 2, 4, 7, 11, ... A comparison with n_e suggests that we write the sequence as $0 + 1$, $1 + 1$, $3 + 1$, $6 + 1$, $10 + 1$, ... In other words, n_s consists of the triangular numbers plus one:

$$n_s = \frac{a_n(a_n - 1)}{2} + 1. \tag{3}$$

Testing our conjectures for $a_n = 8$ and 9, we find that they give the correct result. Thus, we proceed with renewed confidence and seek a way to determine a_n for any given n. If n is a starting position, we can use (3) to find a_n. Specifically,

$$2n_s = a_n(a_n - 1) + 2 \quad \text{and} \quad a_n^2 = a_n + 2(1 - n_s) = 0.$$

Hence by the quadratic formula,

$$a_n = \frac{1 \pm \sqrt{1 - 8 + 8n_s}}{2} = \frac{1 \pm \sqrt{-7 + 8n_s}}{2}. \tag{4}$$

First, we observe that of the two possible solutions, only the one with the positive square root makes sense in this context. Second, if n_s is a value that actually occurs as a starting position, we get the exact answer. A little thought shows that for any n between n_s and n_e inclusive, formula (4) must produce a number whose integer part is the correct value of a_n. Using the usual brackets to denote the greatest-integer, our latest conjecture is that

$$a_n = \left[\frac{1 + \sqrt{-7 + 8n}}{2} \right]. \tag{5}$$

For example, if we want the value of a_6 and a_{25}, we find

$$a_6 = \left[\frac{1 + \sqrt{-7 + 8(6)}}{2} = \frac{1 + \sqrt{41}}{2} \right] = 3$$

and

$$a_{25} = \left[\frac{1 + \sqrt{-7 + 8(25)}}{2} = \frac{1 + \sqrt{193}}{2} \right] = 7.$$

Formula (5) completes our attempt to conjecture a method of determining a_n. But what about the sum $R_n = a_1 + a_2 + a_3 + \cdots + a_n$? Let us again look closely at the series

$$R_n = 1 + 2 + 2 + 3 + 3 + 3 + 4 + 4 + 4 + 4 + 5 + \cdots + a_n.$$

From the structure of the series, one can immediately deduce that if n is such that the series terminates at the end of a string of identical integers, then R_n is just a sum of squares. For example, $R_6 = 1 + 2 + 2 + 3 + 3 + 3$ $= 1^2 + 2^2 + 3^2 = S_3^{(2)}$, and $R_{10} = 1 + 2 + 2 + 3 + 3 + 3 + 4 + 4 + 4 + 4$ $= 1^2 + 2^2 + 3^2 + 4^2 = S_4^{(2)}$.

How do we know when we are at the end of a string? We can find out by using equation (2) and the greatest-integer function, as we used equation (3).

E8: Starting with equation (2), derive an expression similar to (5). Find the value of a_n for $n = 1, 3, 6, 10, 15, \ldots$. What does the greatest-integer form of your expression give if n is not a triangular number?

E9: Write a formula for R_n if n is the last position of a string. What if n is not the end of a string? As you have seen, the formula derived in E8 gives $a_n - 1$ in this case. Then

$$R_n = 1^2 + 2^2 + 3^2 + \cdots + (a_n - 1)^2 + ka_n,$$

where k is an integer between 0 and $a_n - 1$. For example, $R_{12} = 1^2 + 2^2 + 3^2 + 4^2 + (2 \times 5)$, and so $k = 2$.

E10: How is k to be determined? (Hint: If n is a triangular number 1, 3, 6, 10, 15, $\ldots$, then $k = 0$. More generally, k must be the difference between n and the largest triangular number less than or equal to n. Derive an algebraic expression for this statement.)

E11: Another sequence in which the nth term is not easily written down is

$$1, 1, 2, 1, 2, 3, 1, 2, 3, 4, 1, 2, 3, 4, 5, 1, 2, \ldots.$$

Find a formula for the nth term of this sequence. (Hint: Start by writing down the positions in which 1 occurs; then the positions in which 2 occurs; and so on. Try to develop an expression that tells you in which positions the number n occurs. Then the procedure is similar to that used in deriving (5) from (3).)

E12: Find the nth term of the sequence

$$1, 2, 2, 3, 3, 4, 4, 4, 5, 5, 5, 6, 6, 6, 7, 7, 7, 7, 8, 8, 8, 8, 9, 9, 9, 9, 10, 10, 10, 10, \ldots.$$

An Elementary Proof for a Famous Counting Problem

James O. Friel, *California State University, Fullerton, CA*

Most of us try in our teaching to get the student to recognize patterns. In addition to many of the standard elementary patterns, we should all have a few good examples where the "pattern" is not what it appears so that the student will acquire a certain amount of skepticism.

One of the most interesting problems of the foregoing type is the following. Select n points on a circle and form the chords connecting these points (see, e.g., Moise [2], p. 6). How many regions are formed? If one leads the students by showing them that for 2, 3, 4, and 5 points we get respectively 2, 4, 8, and 16 regions, then they will be quite pleased with themselves to respond "32" when asked how many regions will be formed when there are 6 points. The correct answer, however, is that there are at most 31 regions! This could come as a surprise to them. The better students will want to know if there is a general formula and what it is. Fortunately, the answer can be stated in quite simple terms, connecting with—what else? —Pascal's triangle. We propose a proof which requires no advanced techniques.

Before considering the solution to the above problem, consider a related problem. Suppose we have a convex polygon of n sides. What is the maximum number of regions formed by the diagonals? This problem is considered by Honsberger in his excellent little book [1]. He gives three solutions, the nicest of which is included here for completeness. (The reason for asking for the maximum number is that three diagonals may meet in a point.) The process is one of reduction. (See Figure 1. Note that not all diagonals are drawn.)

We investigate the number of regions lost as we remove the diagonal AB. Think of peeling the diagonal off. As AP is taken up, regions 1 and 2 become one region, as PQ is taken up, regions 3 and 4 become one region and so on for every section down the diagonal. Hence the number of regions lost by removing this diagonal is equal to the number of subsegments formed by the points of intersections, which is one more than the number of points of intersection with the other diagonals. Now, removing

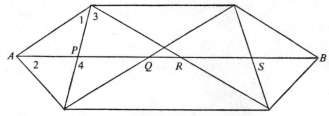

Figure 1.

one of the diagonals also causes its points of intersection with the other diagonals to be removed as well. Thus, as we go from removing one diagonal to another, we have a different number of intersections to consider. Fortunately, in keeping track of the removal process, each point of intersection figures only once. Hence, the number of regions lost in this process of successively removing diagonals is the number of intersections on the 1st diagonal + 1, plus the number of remaining intersections on the 2nd diagonal + 1, plus the number of remaining intersections on the 3rd diagonal + 1, plus , . . . , plus the number of intersections on the last diagonal + 1. The number of terms in this sum is, of course, the number of diagonals. Since each point of intersection is counted exactly once, the value of this sum is the number of intersections plus the number of diagonals. The number of diagonals is $\binom{n}{2} - n$ since of the $\binom{n}{2}$ possible ways of joining n points, n of these are sides of the polygon. The number of points of intersection is $\binom{n}{4}$ since four points determine two diagonals, which in turn determine one point of intersection. Therefore, the total number of regions lost is $\binom{n}{4} + \binom{n}{2} - n$. Now, after removing all the diagonals there is still one region left, i.e., the interior of the n-gon. Consequently, the number of regions is

$$\binom{n}{4} + \binom{n}{2} - n + 1.$$

Note that $\binom{n}{m} = 0$ for $m > n$. Thus, a triangle has exactly $0 + \binom{3}{2} - 3 + 1 = 1$ interior design.

Now let us return to the original problem. We have $n > 0$ distinct points chosen on a circle and wish to count the number of regions formed by connecting these points (Figure 2). The number of regions is clearly the number of regions inside the inscribed polygon plus n, and is therefore $\binom{n}{4} + \binom{n}{2} + 1$. This can be written as

$$\binom{n}{4} + \binom{n}{2} + \binom{n}{0}.$$

Using the fact that $\binom{n}{k} = \binom{n-1}{k} + \binom{n-1}{k-1}$, we can write the above as

$$\binom{n-1}{0} + \binom{n-1}{1} + \binom{n-1}{2} + \binom{n-1}{3} + \binom{n-1}{4},$$

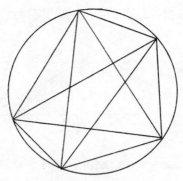

Figure 2.

that is, the sum of the first five elements in the $(n - 1)$th row of Pascal's triangle. (This triangle begins with the 0th one.) The reason why 31 was obtained instead of 32 for $n = 6$ is now clear, and the general pattern is revealed.

REFERENCES

1. R. Honsberger, Mathematical Gems, The Mathematical Association of America, Washington, D.C., 1973.
2. E. Moise and F. Downs, Jr., Geometry, Addison-Wesley, Reading, Mass., 1964.

Problem Solving Versus Answer Finding

John Staib, *Drexel University, Philadelphia, PA*

Frequently, authors of math texts—and teachers who follow them too closely—unwittingly deny students the excitement inherent in problem solving. The solution presented is invariably a textbook-perfect solution: no searching for a strategy, no false starts, and no clumsy algebra. Everything is done neatly and efficiently, giving the impression that solutions to problems are cranked out mechanically and marvelously. Such a presentation, besides being intimidating, gives a very false view of how mathematics is done in practice; it conceals the creative thought process—the spark— that lies behind every nontrivial problem-solving experience.

One way to counter this "polished" approach is to spend more time trying to draw a solution out of a problem presented in class. And if a particular suggestion does not tie in with our preconceived solution, we should nevertheless make an effort to hear the student out and, if possible, to build a solution on the suggestion offered. It may all end in failure. Or, even if successful, the solution may require a long, tedious calculation. In the first case, we may learn from our failure by obtaining a better understanding of exactly why the method failed. In the second case, we can be gratified by our success and yet be dissatisfied by our clumsiness. This may move us to try again, this time aiming for a more elegant solution. All this certainly beats the dubious practice of handing your students a formula and then pretending that they are doing something when they grind out an answer.

Many elementary problems can be used for producing the kind of classroom activity suggested by the above. Here we illustrate one such problem-solving activity. There are ten different precalculus solutions developed here for finding the distance from a point to a line. Some are from individual students, some evolved from the class at large, and some are mine. Each solution, except perhaps for (4) and (10), can be modified to handle the 3-dimensional case. These solutions show how fruitful a single problem can be if we are only willing to cover up the textbook solution.

Problem. *Find the distance from the point* $(2, 3)$ *to the line* $y = x - 1$. (*This particular example is chosen to simplify the arithmetic.*)

Solution 1. (This well-known solution is included because it is one of the easiest to draw from the class.) First, find the equation of the line through $(2, 3)$ that is perpendicular to $y = x - 1$. Next, find the point of intersection of these two lines, and then find the distance from the point to $(2, 3)$.

Solution 2. We introduce a "moving point," say $(t, t - 1)$, on the line $y = x - 1$, and let u be the square of the distance from $(t, t - 1)$ to $(2, 3)$. (See Figure 1.) Thus,

$$u = (t - 2)^2 + (t - 4)^2$$
$$= 2t^2 - 12t + 20.$$

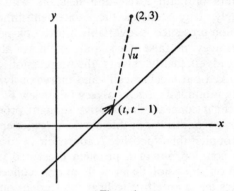

Figure 1.

Now $(t, t - 1)$ will be closest to $(2, 3)$ when u takes on its smallest value. To determine this smallest value, we complete the square:

$$u = 2(t^2 - 6t + 9) + (20 - 18)$$
$$= 2(t - 3)^2 + 2.$$

It follows that u's smallest value is 2 and the required distance is $\sqrt{2}$.

Solution 3. Again we use the "moving-point" fiction of Figure 1. Let v be the slope of the pivoting line segment joining $(2, 3)$ to $(t, t - 1)$. Thus,

$$v = \frac{t - 4}{t - 2}.$$

Now, $(t, t - 1)$ will be nearest to $(2, 3)$ when this pivoting line segment is perpendicular to $y = x - 1$, that is, when $v = -1$. But then

$$\frac{t - 4}{t - 2} = -1.$$

It follows that $t = 3$ and that the nearest point is $(3, 2)$. Finally, the distance from $(3, 2)$ to $(2, 3)$ is $\sqrt{2}$.

Solution 4. Consider Figure 2, where we have introduced certain construction lines to obtain a pair of similar right triangles.

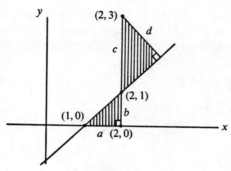

Figure 2.

All labeled points on the figure are readily obtained. Thus, we may write

$$\frac{d}{c} = \frac{a}{\sqrt{a^2 + b^2}}.$$

Or,

$$\frac{d}{2} = \frac{1}{\sqrt{1^2 + 1^2}}.$$

Therefore, $d = \sqrt{2}$.

Solution 5. Consider Figure 3, where a right triangle has been introduced. From the figure, we have

$$d = b \sin \alpha$$
$$= 2 \sin \alpha.$$

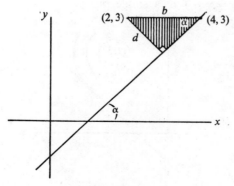

Figure 3.

But $\tan \alpha$ equals the slope of the given line. Thus, $\tan \alpha = 1$ and $\alpha = 45°$. It follows that $\sin \alpha = 1/\sqrt{2}$ and $d = \sqrt{2}$.

Solution 6. Consider Figure 4, where we have again introduced a right triangle. (The vertex $(0, -1)$ could have been any easy-to-find point of the line.) We label the closest point $(x, x - 1)$, and then apply the Pythagorean Theorem:

$$[2^2 + 4^2] = \left[(x - 2)^2 + (x - 4)^2\right] + \left[x^2 + x^2\right]$$
$$4x^2 - 12x = 0$$
$$x = 3.$$

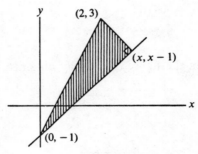

Figure 4.

Thus, the nearest point is $(3, 2)$ and the required distance is $\sqrt{2}$. (The root $x = 0$ is extraneous. Why does it appear?)

Solution 7. Consider a "growing circle" centered at $(2, 3)$, say

$$(x - 2)^2 + (y - 3)^2 = u^2.$$

(See Figure 5.) We let it grow until it is tangent to $y = x - 1$. At that radius it will intersect $y = x - 1$ in a single point.

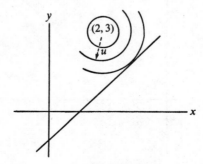

Figure 5.

For a certain value of u, the elimination of y (between our line and the growing circle) will lead to an equation in x that has only a single root. Pursuing this thought, we eliminate y:

$$(x - 2)^2 + (x - 4)^2 = u^2,$$
$$2x^2 - 12x + 20 = u^2,$$
$$x^2 - 6x + \tfrac{1}{2}(20 - u^2) = 0.$$

This last equation will have a single root only if

$$\tfrac{1}{2}(20 - u^2) = (-3)^2.$$

But then $u^2 = 2$. It follows that the radius of the tangent circle—and the required distance—is given by $u = \sqrt{2}$.

Solution 8.   A variation on Solution 7, and a more elementary solution, is the following: Let the "moving circle" grow beyond the tangent point and then "freeze" it. Find the two points of intersection with the line, and then compute their midpoint to obtain the point on the line nearest to $(2, 3)$.

This strategy works best when the circle is made to pass through $(0, -1)$. This circle (Figure 6) has the equation

$$(x - 2)^2 + (y - 3)^2 + 2^2 + 4^2.$$

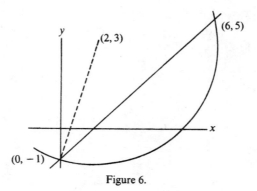

Figure 6.

Eliminating y and simplifying, we obtain

$$x(x - 6) = 0.$$

(That one root would be $x = 0$ was, of course, guaranteed by our choice of radius.) Thus, the points of intersection are $(0, -1)$ and $(6, 5)$, and the midpoint is $(3, 2)$. As before, the distance from $(3, 2)$ to $(2, 3)$ is $\sqrt{2}$.

Solution 9. Consider Figure 7. If we knew the area A of the shaded triangle and its base b, we could obtain d from $A = \frac{1}{2}bd$.

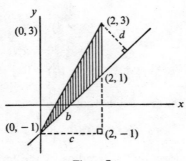

Figure 7.

All points labeled on the figure are readily obtained. It follows directly that $b = \sqrt{8}$. To find A, we take the difference of the areas of the two right triangles having base c:

$$A = \tfrac{1}{2}(2)(4) - \tfrac{1}{2}(2)(2) = 2.$$

Therefore,

$$d = \frac{2A}{b} = \frac{4}{\sqrt{8}} = \sqrt{2}.$$

Solution 10. (This solution is a bit more technical, requiring knowledge of the rotation of coordinate systems.) Consider Figure 8, where we have raised the point and line by 1 unit and then introduced an $x'y'$-coordinate system.

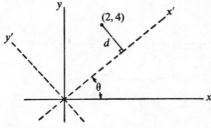

Figure 8.

The required distance d is now seen to be the y'-coordinate of the point $(2,4)$. Thus, using the standard formulas of rotation, we may write

$$d = y' = -x \sin\theta + y \cos\theta$$
$$= -2 \sin\theta + 4 \cos\theta.$$

But $\sin\theta = \cos\theta = \sqrt{2}/2$ can be deduced from the observation that θ is a first-quadrant angle and that $\tan\theta = 1$. So,

$$d = -2\frac{\sqrt{2}}{2} + 4\frac{\sqrt{2}}{2} = \sqrt{2}.$$

A closing tale. One day, after the class and its teacher had solved the above problem in three different ways, a student jumped up in an agitated state and said, "What are we doing here? We got the answer to that problem 15 minutes ago and yet you and some others keep on talking about it and looking for other ways to get the same answer over again. Why?"

The teacher, delighted by this opening, responded as follows: "There is a great difference between solving problems in a business or industrial setting and doing the same thing in an educational setting. In the first instance, problems are solved because someone needs the answers; indeed, these answers have economic value. And since "time is money," anything goes: we use handbooks of formulas, tables, computers, solutions to similar problems that are on file, trial and error, physical models, and so on. But here in the classroom our goals are educational rather than economic. This means that our interest in problem solving is not primarily to get an answer; we solve problems in order to gain experience in the art of problem solving, and thus acquire a deeper appreciation of how a fragmentary set of purely mathematical ideas can be organized into a powerful strategy for solving a problem. The real delight of mathematics is to solve a problem using our wits and general knowledge of mathematics, rather than by using a formula or other answer-finding recipe. Since today's lesson has been devoted to *problem-solving*, rather than *answer-finding*, it is entirely appropriate that we should attempt to solve the same problem in many different ways."

Using Monte Carlo Methods to Solve Probability Problems

Carolyn Shevokas Funk, *Thornton Community College, South Holland, IL*

College mathematics textbooks generally emphasize analytic methods for solving probability problems. Students first find probabilities by considering the total sample space (the set of all possible outcomes). Then they are introduced to the addition rule for mutually exclusive events and the multiplication rule for independent events. Although students usually appear interested in the simple textbook problems, many have little success in using these analytic methods to obtain solutions. There is another effective alternative.

Monte Carlo methods [7] involve expressing a problem situation as an experiment in probability, and then performing the experiment a number of times in order to obtain an estimate of the solution to the problem. Monte Carlo methods have usually been used in situations where an exact mathematical solution is too difficult to obtain or simply does not exist, such as in finding solutions to Boltzmann equations [2]. The Monte Carlo method, as used for the study described in this article, was applied to problems in which an exact mathematical solution exists but was beyond the easy grasp of the students involved. Thus, it was used for pedagogical rather than mathematical reasons.

Example 1. In a family of three children, what is the probability that all three will be boys?

Monte Carlo Solution. We can express our problem as a coin-tossing experiment. It is assumed that boys and girls have equal chances of being born.

1. Let heads represent "boy is born" and tails "girl is born."
2. Toss three coins in succession (to represent a family of three children).
3. A success occurs when all three tosses show heads.
4. Repeat the above steps many times.
5. The ratio of the number of successes to the number of trials gives an estimate of the desired probability.

If in 50 tosses of three coins all heads were obtained five times, the probability estimate of having three boys in a family of three would be 5/50 or 0.1.

Each of 30 students performed fifty trials. The following is a summary of the estimates they obtained:

Probability Estimate	.04	.06	.08	.10	.12	.14	.16	.18	.20
Frequency	2	2	3	5	4	6	3	4	1

The results of all 30 students (1500 trials) yielded a mean value of 0.122 with a standard deviation of 0.0434.

Analytic Solution. Using the formula for independent events, the exact solution to the problem is found to be P(three boys) $= (1/2) \times (1/2) \times (1/2) = 1/8 = 0.125$. Note that all students obtained results within .085 of the exact solution 0.125. The estimated mean (0.122) for all 1500 trials is within 0.003 of this true value.

It can be shown [5] that 168 trials are needed to be 95% certain that the estimate to this problem is within 0.05 of the true value 0.125. To be 99% certain that the estimated value is within 0.05 of the true value, about 300 trials are needed.

Example 2. A manufacturer of cameras is having difficulty because 10% of the springs he uses are defective. In the next lot of five springs that he uses, he wants to know what the probability is that one or more will be defective.

Monte Carlo Solution. Consider using one-digit random numbers.

1. Let the digit "0" represent the event "defective spring," and let the digits "1, 2, . . . , 9" be the event "non-defective spring."
2. Read off five 1-digit numbers from a table of random numbers.
3. A success occurs whenever one or more of the five digits represent defective springs (i.e., whenever at least one of the five digits is "0").
4. Repeat the above steps many times.
5. The ratio of the number of successes to the number of trials gives an estimate of the desired probability.

If $(4, 3, 0, 1, 7)$ is a series of one-digit numbers obtained from a random number table, then the third spring would be considered defective and the trial would be a success. Fifty trials were performed by using a table of random numbers, and 19 were found to be successes. Accordingly, the estimated probability of having one or more springs defective in a lot of five was $19/50 = 0.38$.

Analytic Solution. Using the formula for independent events,

$$P(\text{no defectives in five}) = (.9)^5 = 0.59049,$$

we obtain the exact solution

$$P(\text{at least one defective in five}) = 1 - 0.59049 = 0.4095.$$

Monte Carlo Methods in the Classroom

Monte Carlo methods for solving problems can be particularly useful to students without much demonstrated ability in mathematics since there is little need for theorems and complex notation. This is especially true since analytic methods become more complex as problems become more difficult, but Monte Carlo techniques really do not.

"What, if anything, would be the effect on students' problem solving if they were taught only Monte Carlo techniques as compared with being taught only conventional analytic methods?"

Purpose and Description of the Study. The objectives of this study were to determine whether or not:

(1) Monte Carlo techniques would help students, at this level, learn to solve probability problems with greater success than would similar students using conventional analytic methods.
(2) The computer could be effectively used with Monte Carlo methods by students having little, if any, previous computer experience.
(3) There would be an improved attitude toward the study of probability (and toward mathematics in general) for students using Monte Carlo techniques as contrasted with students using only analytic techniques.

Seventy-one students at Polk Community College in Winter Haven, Florida, were divided into two experimental groups and one control group. Students in both experimental groups learned to solve probability problems only by using Monte Carlo techniques. Students in the control group learned to solve probability problems by using only analytic methods. The distinction between the two experimental groups was that one of them had access to the computer and the other did not. Students in the noncomputer experimental group had to do all of their problems by hand methods that involved using such physical objects as coins, dice, cards, and tables of random numbers; those in the computer experimental group had the choice between hand and computer methods. (Materials for use in both experimental groups were identical, except for the reference to computer usage in the materials for students in the computer group. A library of computer programs in BASIC for solving problems by Monte Carlo procedures was available to students in the computer-oriented group.)

Results of the Study

(1) All groups were administered achievement tests at the beginning and end of the experimental period. On the pre-experiment achievement test (the Cooperative Arithmetic Test), there was no significant difference between group means. The post-experiment achievement test* showed a significant difference ($p < 0.01$) between group means, in favor of the two experimental groups; there was no significant difference between the experimental groups themselves.

(2) Pre- and post-measures of attitude toward mathematics were obtained by using both the Aiken-Dreger Scale [1] and the McCallon-Brown Scale [4]. No significant differences between sample means were found on these attitude measures for the pre-experiment scores. The same was true for the post-experimental scores. A t-test analysis of the change in attitude toward mathematics during the study of the probability unit, as reflected by pre- and post-scores on the attitude measures, showed a significant positive gain in attitude ($p < 0.05$) toward mathematics for students in the noncomputer group; a slight significant deterioration in attitude ($p < 0.05$) for students in the control group; and a slight nonsignificant negative change for students in the computer experimental group.

(3) Students in all groups responded to a questionnaire to help determine how they felt about their study. Significant differences in responses indicated that students in the Monte Carlo groups found probability more interesting and more fun than did students in the control group. Furthermore, students in the Monte Carlo experimental groups felt that problems were more fun to solve and not as difficult (though many were more difficult than the problems encountered by students in the control group).

General Remarks. Students in the experimental classes generally appeared much more enthusiastic during class than those in the control group; they responded more, made more suggestions, and seemed to be more involved. Homework discussions in the control group were very teacher-oriented, despite the fact that students were continually being urged to make suggestions and contributions. In the experimental groups, homework discussions became much more student-centered since these students were eager to share their ideas (especially when they felt their ideas were original), and they did not seem to be as fearful of being wrong as did students in the control group.

---

*Students in the experimental groups were required to describe correct Monte Carlo procedures and obtain results using at least ten trials in order to receive credit for a problem. Control group students received credit for correct numerical solutions. No partial credit was given on this test.

After the study of probability was completed, students in all groups studied number bases, sets, abstract systems and logic. Students who had been in the experimental Monte Carlo groups appeared more enthusiastic throughout the remainder of the course and they were more willing to share their ideas in class. One possible inference is that success with the Monte Carlo unit gave many of them more confidence in their ability to handle mathematics.

The computer, we see, was not indispensable for using Monte Carlo methods in classroom situations. In fact, 20–25 trials are usually enough for a reasonably accurate estimate and this number of trials can be performed quickly by hand methods. The computer's use presents benefits as well as obstacles. Students in the noncomputer group seemed to improve their attitude toward mathematics more than did students in the computer group. On the other hand, setting up a procedure for using the computer programs required more abstraction than was required for hand methods and this places more strain on the poorer students.

Conclusion. As a result of this experiment, I feel that Monte Carlo methods can be used successfully by students having limited ability and interest in mathematics—students who might otherwise have little success using analytic methods to solve problems in probability. More capable students should be encouraged to learn both methods. For them, Monte Carlo methods can be used to verify analytic results as well as to solve problems too difficult to be handled analytically.

All students can benefit from the model-building experience formulated in Monte Carlo techniques.

REFERENCES

1. Allen L. Edwards, Techniques of Attitude Scale Construction, Appleton-Century-Crofts, New York, 1957.
2. J. M. Hammersley and D. C. Handscomb, Monte Carlo Methods, Methuen & Company, London, 1964.
3. Allen H. Holmes, Teaching the logic of statistical analysis by the Monte Carlo approach, Ph.D. dissertation, University of Illinois, 1968.
4. Earl L. McCallon and John D. Brown, A semantic differential instrument for measuring attitude toward mathematics, J. Experimental Education, 39 (1971) 69–79.
5. Carolyn Shevokas, Using a computer-oriented Monte Carlo approach to teach probability and statistics in a community college general mathematics course, Ph.D. dissertation, University of Illinois, 1974.
6. Julian L. Simon and Allen H. Holmes, A really new way to teach probability statistics, Math. Teacher, 62 (April 1969) 283–288.
7. I. M. Sobol, The Monte Carlo Method, MIR Publishers, Moscow, 1975.

Introducing the Computer in Liberal Arts Mathematics

S. P. Gordon, R. W. Meyer, and A. Shindhelm, *Suffolk Community College, Selden, NY*

1. Introduction. Amidst predictions of computers in every home, the mathematics community is reacting by gradually incorporating computer usage into many of its courses. Several studies (see [2] for example) indicate that the computer is used most often in the middle level math courses such as calculus, differential equations and numerical analysis. Less frequently, it is used as a remediation tool using the CAI (computer assisted instruction) mode. Very rarely is it used in the broad based liberal arts level courses which are often our "bread and butter" courses. However, the authors' experiences indicate that the computer can play an important and constructive role in courses at this level. Moreover, these courses present a wonderful opportunity to expose a large audience to the computer in a variety of ways which are educationally useful and extemely enjoyable to all concerned.

As described in a recent article [1], Suffolk Community College has been awarded a quarter million dollar grant under the NSF's Comprehensive Assistance to Undergraduate Science Education (CAUSE) program to develop and implement a program to incorporate computers into a wide variety of courses. The academic areas covered by the grant include mathematics, the physical sciences and engineering, and economics. The present paper discusses some of the developments under this project in the mathematics areas with particular emphasis on the liberal arts math courses.

The principal objective of the CAUSE project at Suffolk is to expose as many students as possible to the computer, both as an educational aid in learning the concepts and techniques in a course and as a practical tool for solving the types of sophisticated problems which are usually much too complicated to consider in class. However, in order to have any sizable impact on the students, it is first necessary to involve as many faculty members as possible. Thus, the primary thrust of the project is faculty development in programming and in the educational uses of computers.

Essentially, the grant provides the funds for two types of interrelated activities.

(a) Selected instructors are offered a modest amount of released time during which they prepare packages of computer programs, with accompanying student and instructor manuals, designed for particular courses.

(b) We are offering a number of summer workshops, using BASIC language, to train the faculty in programming. At the same time, we familiarize them with the packages and related materials so they can easily incorporate the computer into their courses.

About 40 faculty participated in the first workshops offered during a three-week session during the summer. Of these, about 10 were from the mathematics department. Approximately 10 other math instructors became involved in the use of the computer through developing packages under the grant or through smaller scale use of the computer. Consequently, 70% of the college's mathematics faculty are currently using the computer.

At present, the first four packages have been developed; the last three packages should be completed shortly:

1. Traditional calculus sequence—S. P. Gordon
2. Survey of mathematics—A. Shindhelm
3. Finite mathematics—D. Coscia
4. Descriptive and inferential statistics—R. W. Meyer
5. Tutorial in remedial arithmetic—P. Herron
6. Elementary functions—T. Hasiotis
7. Graphics for calculus—S. P. Gordon

We anticipate that the greatest impact in the mathematics department, in terms of the number of students affected, will be in the liberal arts math offerings which consist of two semesters of survey of mathematics and a one-semester introductory statistics course. There is also a follow-up statistics course that extends the first course and involves students in the preparation, administration and analysis of an actual survey.

2. Survey of Mathematics Sequence.

In the survey of math courses, the computer is being used in two ways. One approach treats the computer as a unit by first teaching some elementary programming techniques in BASIC, and then assigning individual computer projects to the class. Another approach is to use the computer to enhance topics taught in these courses.

A. Shindhelm's programs use both the computational and simulation modes of the computer. Thus, the computer's built-in random number generator is utilized to illustrate concepts of probability theory. For example, programs have been developed which simulate the tossing of a coin N times and the tossing of three coins N times. In both cases, the student chooses the number of trials N and the computer responds with an analysis of the simulated results. By displaying the numerical results of such experiments, the computer can provide students with additional insight into just what the probabilities of the various outcomes represent and how the results of the experiment match the predicted values based on the theory.

Number theory and group theory are two other topics in the survey of math sequence. The computer's use in these topics simplifies the students' involved computations and therefore better enables them to appreciate the concepts involved. Thus, for example, there is a program that determines the greatest common divisor of two integers and prints out the details of the Euclidean Algorithm used in its computation. Another program illustrates the Chinese Remainder Theorem by solving simultaneous congruences. Still another program of this type checks for commutativity and associativity of any binary operation that is entered.

A series of programs has also been developed which can be used for classroom demonstration, as well as for students to check their homework problems. For instance, one of the programs for studying symbolic logic prints out the truth table for any logical expression which is entered, while another program determines if any two logical expressions entered are logically equivalent. Programs dealing with the topic of number bases are also available. One program converts a number from one base into an equivalent number in another base. Another program illustrates the four basic arithmetic operations of addition, subtraction, multiplication and division in any base.

It is important to consider the relatively nonmathematical orientation of the students taking this survey type course. In this vein, all of the above programs have been designed for use by individuals with virtually no knowlege of computers and programming. Only a few minutes are required in class to demonstrate how to turn on a terminal and how to key in the few necessary characters. As mentioned above, the computer has also been used in these courses as an independent unit. The computer projects assigned to the class are chosen to match the previous mathematical backgrounds of the students. Thus the project for students with minimum elementary algebra/geometry prerequisite involves having the computer calculate areas and perimeters for simple geometric figures (circles and parallelograms). For those students who have had intermediate algebra, the project involves calculating the roots of a quadratic equation. For those with trigonometry, the project is to make use of appropriate identities to have the computer calculate the values of each of the six trig functions. For those students who have completed the equivalent of 12th year mathematics, the project is to find the roots of any arbitrary function (usually a polynomial) using the bisection method. For those who have had statistics, the project involves calculating the mean and standard deviation for a set of numbers.

It has been found that a total of three to four weeks, or approximately 25% of the course, is usually required to cover the computer unit. It is necessary to spend several hours on flowcharting and about two hours on explaining the rudiments of the BASIC language. The balance of the time is devoted to working with students on a one-to-one basis to guide and/or push them through the problem's logic and the programming pitfalls. The end results, described later, make this experience very worthwhile for students.

3. Statistics Sequence. In the introductory statistics course, the computer is used via "canned" programs. The programs, developed by R. W. Meyer, are designed to require no programming skills by the students using them. These programs can be used either as a computational aid (in or out of the classroom) or they can be used in the classroom to demonstrate the power and sensitivity of statistical calculations. In the latter case, for example, one of the programs is a three-level program which calculates the mean and standard deviation of user data; draws a random or biased sample from either user data, uniform data or normally distributed data; and demonstrates the Central Limit Theorem for means. Another program explores the effect of changes in data on a descriptive statistic. That is, it illustrates how the standard deviation reacts to shifts in data ($X \to X + A$), to magnifications ($X \to KX$) and to changes in spread with fixed mean. Still another program illustrates the effects of changes in sample size, data, or level of significance on the final conclusion of a hypothesis test. For example, is a sample with 45% favoring some candidate significantly different from a population with 50% in favor if the sample size is 30, 90, 300 . . . ? Demonstrations of this type are often impossible to do by hand in class (even using calculators) because of the time involved.

It is expected that the students will not only learn statistics more fully through use of these programs, but will also gain some insights into the nature of a computer. In particular, the speed of computation and the need for user precision should become apparent.

4. Evaluation of Results. In evaluating the success of efforts such as those described above, it would be ideal to demonstrate statistically that students in sections using computers did significantly better than students in more traditional sections. However, when we consider the tremendous variety of students who take the types of courses discussed in this paper, it is evident that any such analyses would be practically impossible and any conclusions based on them would be valueless. Let us focus instead on the attitudes of the individuals involved in order to draw conclusions regarding the desirability of incorporating the computer into liberal arts mathematics courses. The following results are drawn from surveys administered to over 160 students who were registered in classes using the computer during the fall 1978 semester.

As described in the preceding sections, the computer was used in several distinct modes in these courses—either for demonstrations of course topics or as an end in itself. Accordingly, we will address the student reactions separately for each type of usage. In evaluating these experiments, we consider student reactions to the three principal objectives for introducing the computer:

1. to enhance students' understanding of the course material;
2. to expose students to the capabilities and uses of the computer;
3. to make the course interesting to students.

Clearly, different instructors would assign varying weights to these objectives depending on the course and individual philosophy. In addition, it is felt that a key determinant to student reactions to the use of the computer is the extent to which they would like to see the computer usage continued in the future.

In A. Shindhelm's survey of math classes, virtually all of the students entered the course with no previous computer background. Sixty-two percent of the students felt that the computer did make the course more interesting while 14% disagreed. In addition, 75% felt that they had developed a better understanding of the computer's capabilities. While 29% felt that the computer had been of assistance in understanding the course materials, 23% felt the computer had not been of any assistance. The remainder seemed unsure as to whether or not the computer had helped them. In summary, 43% felt that the use of the computer should be increased, while 50% felt the present level of usage was fine. As if to punctuate this, the overwhelming majority of written comments indicated that many of the students believed that greater interplay with the computer would have been extremely desirable and beneficial.

In R. W. Meyer's statistics classes, the students apparently had a somewhat greater familiarity with computers, since the majority indicated having had minimal computer experience. In these classes, where the computer was also used on a demonstration basis, the primary objective was to have the computer enhance the students' understanding. Here 56% asserted that the computer had helped them, while 5% felt that it hadn't. The remaining 39% seemed unsure. And although 36% of the students believed that they had gained some appreciation of the uses of the computer, 25% disagreed with this statement. Finally, the students were fairly evenly split as to whether the role of the computer should be increased or continued at the present level. Only 6% felt that its use should be decreased.

In S. P. Gordon's survey of math classes, virtually none of the students had had any previous computer experience. The primary objective of introducing the computer was to give students an appreciation of its capabilities, and 96% agreed that this had been achieved. Here 89% found that the computer unit had made the course more interesting. The computer had also been used for demonstrations on a small scale basis. In this regard, 33% of the students believed that it had helped their comprehension of the material, while 7% felt it was not useful at all. As intimated in an earlier section, in order to complete their computer projects, the students had to put in more time and effort than for any other aspect of the course. Nonetheless, 33% indicated that they would prefer even more computer usage in the course, whereas 64% felt it should be continued at the present level. Only one student felt it should be decreased.

In conclusion, it is clear that the overwhelming majority of students involved welcomed the inclusion of the computer into their courses regardless of the mode of usage. Equally importantly, the instructors feel that the computer can play an important and constructive role in courses such as these. Liberal arts courses provide a unique opportunity to expose large student audiences, to the computer, which might otherwise have absolutely no direct contact or experience with computers. The ways in which the computer can be employed are educationally beneficial and very enjoyable to both students and instructors.

REFERENCES
1. S. P. Gordon, Development of a computer-oriented program in science and mathematics, Proc. of Ninth Conf. on Computers in the Undergraduate Curricula, University of Denver (1978) 344–351.
2. James W. Johnson, The State of the Art of Instructional Computing, A report from CONDUIT (to appear).

Mathematics Placement Test Construction

F. Pedersen, *Southern Illinois University, Carbondale, IL*

Introduction. Having been asked to handle the placement and proficiency program in mathematics at SIU, I quickly discovered that my knowledge relevant to this area of counseling was limited, and further investigation showed me that my colleagues were no more informed on this topic than I was. After having discussed this problem with other mathematicians on other campuses, I realized that this was a typical situation.

Most schools were either using commercially developed placement tests, or else they were designing their own placement tests. The people designing their own placement tests were not always sure that the test constructed was a reliable measure of present knowledge or reliable predictor of future success.

Realizing that I was not going to get very much help from outside sources and that the placement test had to be constructed, I chose a plan of attack and proceeded to construct a test. The test had to be taken by the student in no more than 90 minutes, had to be composed of multiple choice questions for easy grading by a nonmathematically trained person, and had to differentiate between five levels of mathematical competency. The method of test construction was also to be designed to utilize the computer in evaluating the reliability of the questions as well as in determining the pass/fail cut-off scores. Decisions based on statistical data, rather than on my subjective evaluations, were to increase the effectiveness of the constructed test.

This article discusses some of the practical aspects of test design that were formalized while constructing this test. The effective use of a computer in processing test responses, as well as the results of follow-up studies concerning the reliability of the constructed placement exam, will also be considered.

The process of test construction described can also be used by mathematicians who may not have extensive training in computer use and test construction design.

Basic Philosophy. The following principles were adopted for the efficient construction of multiple choice test problems.

1. Do not begin to construct test questions until there is an outline of the topics that are considered to be basic knowledge in the area to be tested.

2. Design simple straightforward questions. Avoid the desire to construct clever questions since the object is to determine the student's content knowledge, not creative-aptitude.

3. Vary the difficulty of the questions from very easy to hard. Always start each area to be tested with a few easy questions.

4. Do not use "None of the above" as a possible answer since this choice will adversely affect better students more often than poorer students. (A slight error will frequently cause the good students to choose "None of the above" without any further check of their calculations.) One can always construct plausible test answer choices by making typical mistakes in working out the problems.

5. Avoid asking for technical information which is not likely to be remembered by the student. In general, students do not prepare extensively before taking placement tests.

6. It is not necessary to ask questions covering every topic in the area to be tested. Some principles are so closely related to each other that knowlege in one area almost invariably implies knowledge in the related area. The effectiveness of the test depends a lot more on the quality of the questions than on the total coverage of the subject area.

Computer Evaluation of Questions. Having constructed several possible test questions and having administered the questions to a well-defined group of students, one needs some method of determining if a question is a valid measure of the students' abilities. The computer serves as a data-gathering tool for making this decision. In this paper, we used a computer print-out called an *ANLITH* item analysis. There are only a few facts one needs to understand about the *ANLITH* print-out.* The "computer analysis" found with each of the sample questions in the next section is an example of an *ANLITH* item analysis. It provides data, relating the effectiveness of an individual question to the type of student tested, in the following manner:

1. The test data is presented with reference to the performance of five groups of students who took the test. These groups of students, determined by the Z-score cutoff points, will each contain approximately 20% of the students if the data is normally distributed. The distribution of Z-scores, from negative to positive, is included in the computer analysis of the test as

---

*My personal thanks are extended to Dr. Gordon White of the SIU-C Counseling and Testing Center for his help with all of the computer analysis used.

a whole. The chart (Table 1) indicates that Group 1 includes all students who had a Z-score below -0.84, and that this was 15% of the total students (or 12 students). The information on Groups 2 to 5 is interpreted similarly. It should be noted that the students in Group 1 received the lowest scores on the test; the students in Group 2 were the next highest scorers; and so on up to Group 5, which represent the students who scored the highest.

Table 1.

Group	Z-cutoff	Percent in Total Group	No. in Each Group
1	-0.84	15.00	12
2	-0.26	18.75	15
3	0.26	22.50	18
4	0.84	27.50	22
5	6.00	16.25	13

2. The first table given in the (*ANLITH*) computer analysis of a question is called the Frequency Distribution Graph. It indicates how many students from each group chose each of the answers A, B, C, D, or E (1, 2, 3, 4, or 5 respectively). The 0-column indicates the number of students who did not answer the question. The correct response is indicated below the chart.

3. The second table given in the (*ANLITH*) computer analysis is a bar graph illustrating the percentage of students in each group which answered the question correctly. The ideal question would have the end-points of the percentage graphs determining a line with negative slope.*

4. The print-out gives an item-test R number. This is the Pearson product-moment correlation coefficient that correlates the results of the test question with the results of all the questions given in the test. A so-called "perfect" question has a coefficient of 1, but actual results from long-time

---

*Since the five divisions are divided into the lowest 1/5 of total scores, the second lowest 1/5 of the total scores . . . the top 1/5 of the scores on the test, it is clear that a good question would differentiate between these groups of students in exactly this same way. That is, the lowest group should have the least number of correct responses . . . the highest group should have the highest percentage of correct responses. Thus, the graph should resemble:

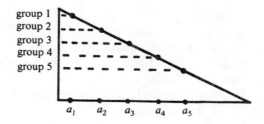

group 1
group 2
group 3
group 4
group 5

a_1 a_2 a_3 a_4 a_5

usage indicate that a good question will have an *R* number above .400 and near to .500.

5. An item difficulty number is given. It is the (weighted) mean percent of students who got the problem correct. This number may also be viewed as the (weighted) mean length of the bars in the percentage bar graph.

The following examples are given to illustrate the *ANLITH* computer analysis. They will be used to discuss how a computer can provide the basis for accepting or rejecting a test question. We will also refer to the basic principles on placement test construction formulated earlier.

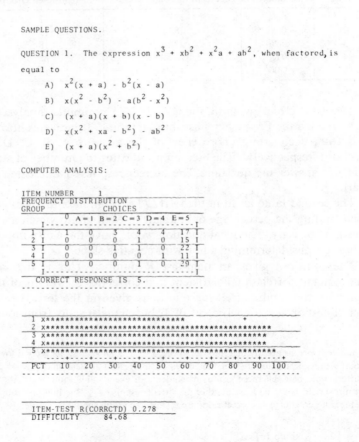

SAMPLE QUESTIONS.

QUESTION 1. The expression $x^3 + xb^2 + x^2a + ab^2$, when factored, is equal to

A) $x^2(x + a) - b^2(x - a)$

B) $x(x^2 - b^2) - a(b^2 - x^2)$

C) $(x + a)(x + b)(x - b)$

D) $x(x^2 + xa - b^2) - ab^2$

E) $(x + a)(x^2 + b^2)$

COMPUTER ANALYSIS:

```
ITEM NUMBER        1
FREQUENCY DISTRIBUTION
GROUP               CHOICES
            0   A=1  B=2  C=3  D=4  E=5
   I------------------------------------I
 1 I    1    0    3    4    4    17  I
 2 I    0    0    0    1    0    15  I
 3 I    0    0    1    1    0    22  I
 4 I    0    0    0    0    1    11  I
 5 I    0    0    0    1    0    29  I
   I------------------------------------I
     CORRECT RESPONSE IS  5.
```

```
 1 X*****************************
 2 X*********************************************
 3 X*******************************************
 4 X****************************************
 5 X************************************************
   ---+----+----+----+----+----+----+----+----+----+
PCT   10   20   30   40   50   60   70   80   90  100
```

```
   ITEM-TEST R(CORRCTD) 0.278
   DIFFICULTY       84.68
```

Question Analysis. Question 1 was clearly an easy question for the students who took the test. Assuming that the test was given to the proper level of students, this question should be used only for a starter question. It is advisable to have several such easy questions at the beginning of the test. Note that answer A deceives no one, and thus should be changed to a plausible solution.

QUESTION 2.

$$\left(1 - \frac{x}{1 - \frac{1}{x}}\right)(1 - x) \quad \text{is equal to}$$

A) $-x^2 + x - 1$

B) $x^2 - x + 1$

C) $x^2 - 1$

D) $x^2 + x + 1$

E) None of the above

COMPUTER ANALYSIS:

ITEM NUMBER 2 TEST NUMBER						
FREQUENCY DISTRIBUTION						
GROUP		CHOICES				
	0	A=1	B=2	C=3	D=4	E=5
1	1	4	5	2	4	13
2	1	3	2	1	0	9
3	2	4	4	1	0	13
4	2	0	4	2	0	4
5	1	1	17	1	1	9

CORRECT RESPONSE IS 2.

```
1 X*********        +
2 X******          +
3 X*********        +
4 X*******************
5 X****************************************
   ---+----+----+----+----+----+----+----+----+----+
PCT  10   20   30   40   50   60   70   80   90  100
```

ITEM-TEST R(CORRCTD) 0.248
DIFFICULTY   28.83

The problem in Question 2 is difficult for an algebra student because there are so many opportunities for errors in solving the problem. This example illustrates how the answer, "None of the above" can be unsuitable; its selection here only shows that all levels of students can make errors.

Question 3 is working very well. There appears to be very little possibility that the students will guess the answer. The graph is almost perfect in the respect that this question clearly differentiates between high and low groups. It is true that you do not want too many questions of this type on a test.*

---

*There is very little difference between the graphs of questions 2 and 3—except for the fact that the bottom three groups could not even guess the right answer on question 3, whereas some from all three lower groups were able to get the correct response to 2. Since the object is placement, this question 3 seems to be doing an excellent job of separating the top two groups from the bottom two groups. On the other hand, question 3 does nothing to help differentiate between the bottom three divisions. A placement test must clearly separate all the students, not just the top 40%.

QUESTION 3. The expression $(\tan \theta + \cot \theta)^2$ is equal to

A) 1

B) $\tan^2\theta + \cot^2\theta$

C) $\sec 2\theta$

D) $2 \tan \theta \cot \theta$

E) $\sec^2\theta + \csc^2\theta$

COMPUTER ANALYSIS:

```
ITEM NUMBER    3                    TEST NUMBER
FREQUENCY DISTRIBUTION
GROUP              CHOICES
        0  A=1  B=2  C=3  D=4  E=5
    I--------------------------------I
  1 I    5    6    5    1    0    0 I
  2 I    1   19   12    2    7    0 I
  3 I    0    5    5    2    3    0 I
  4 I    3    3    2    0    5    2 I
  5 I    2    3    2    2    2   12 I
    I--------------------------------I
    CORRECT RESPONSE IS   5.
```

```
  1 X       +
  2 X·      +
  3 X       +
  4 X*******
  5 X***************************
    ----+----+----+----+----+----+----+----+----+----+
PCT    10   20   30   40   50   60   70   80   90  100
```

```
ITEM-TEST R(CORRCTD) 0.555
DIFFICULTY     12.61
```

Note the Item-Test R-number is .555—above the range usually recommended by testing professionalis. It is, nevertheless, a question which helps to identify the best students. Because the ideal R-number may not be necessary to meet the objectives of a question, the graph and the table are much more indicative of a question's reliability than is the Item-Test R-number.

Question 4 is clearly not providing accurate information about the ability of the students. This may be due to the use of "None of the above" as a possible answer, and/or the unnecessary complication of the statement of the problem by such words as "units." The Item-Test R-number of $-.015$ is consistent with the lack of reliability of the question. It is better to eliminate such poor questions rather than try to improve them.

Question 5 is very good and can be used effectively as a moderately difficult question. While it may not distinguish between the top two groups, it certainly does scale down in a desirable manner. Note that the Item-Test R-number of .283 does not agree with my positive feelings concerning the use of this question.

QUESTION 4. In the picture below, the circle has a radius of 10

units. The arc AB is 8 units long. What is the radian measure

of the central angle, AOB ?

A) 8π/10

B) π/2 - 4/10

C) 8/10

D) 80

E) None of the above.

COMPUTER ANALYSIS:

ITEM NUMBER	4			TEST NUMBER	

FREQUENCY DISTRIBUTION
GROUP   CHOICES

	0	A=1	B=2	C=3	D=4	E=5
I--I						
1 I		5	4	4	0	2	2 I
2 I		4	14	6	5	1	11 I
3 I		0	6	0	3	0	6 I
4 I		1	7	1	2	2	2 I
5 I		3	4	2	2	1	11 I
I--I						

CORRECT RESPONSE IS 3.

```
1 X      +
2 X******
3 X***********
4 X********
5 X*****+
   ----+----+----+----+----+----+----+----+----+----+
PCT  10   20   30   40   50   60   70   80   90  100
```

ITEM-TEST R(CORRCTD) 0.015
DIFFICULTY   10.81

Question 6 is included to illustrate that "None of the above" is not always detrimental to the better student. It really depends on the type of question asked and the types of mistakes expected—careless or lack of knowledge. Note, in the analysis, that answer B is not a very likely choice. So by changing it, the question could possibly be improved.

By utilizing this method of question analysis, aided by computer-organized data, it is possible to reduce the number of test questions to a minimum with confidence that each question will do its intended duty.

Test Construction. This section will briefly describe the mechanics of test construction used at SIU. First, three 25-30 question tests were constructed—one test for each of the three mathematical areas that the students were to be tested: intermediate algebra, college algebra and trigonometry I, and college algebra and trigonometry II. The decision to use 25-30 test questions was based on the fact (determined by trial and

QUESTION 5. The degree measure of the angle with radian measure
equal to 10 is

    A) between 540° and 630°

    B) between 270° and 360°

    C) over 1500°

    D) less than 270°

    E) between 720° and 1080°

COMPUTER ANALYSIS:

```
ITEM NUMBER    5              TEST  NUMBER
FREQUENCY  DISTRIBUTION
GROUP              CHOICES
          0  A=1  B=2  C=3  D=4  E=5
    I--------------------------------I
 1  I    5    1    2    2    6    1 I
 2  I    1   14   11    5    5    5 I
 3  I    0    6    4    2    1    2 I
 4  I    0   10    0    3    1    1 I
 5  I    2   15    0    1    3    2 I
    I--------------------------------I
    CORRECT RESPONSE IS   1.
```

```
 1  x***                    +
 2  x******************     +
 3  x********************* +
 4  x*****************************************
 5  x****************************************
   ----+----+----+----+----+----+----+----+----+----+
PCT   10   20   30   40   50   60   70   80   90  100
```

```
ITEM-TEST R(CORRCTD)  0.283
DIFFICULTY      41.44
```

experience) that a student can answer about that many multiple-choice
questions comfortably in 50 minutes. Secondly, two sections of classes in
each testing area were chosen, and the appropriate test was administered to
these students. At least 50 students are needed for each test to determine a
reliable item analysis. The testing was done during the first week of classes
so that the students would be mathematically similar to students involved
in registration and placement. The third step was to obtain the *ANLITH*
analysis of the test data. Using the item analysis in the manner described in
the previous sections, "good" questions were chosen and a 63-question test
was constructed. Because the actual test was covering three areas of
questions, the test was divided into three sections. It was felt that students
with scores in the same range as Groups 4 and 5 of the sample population
should be in advanced placement. Thus, it was fairly easy to predict cut-off
scores by examining the item analysis data.

    While the exact cut-off scores are of little interest, except for SIU's
internal use, the method of utilizing these scores can benefit other universi-
ties. The cut-off scores decided on were based on the following test

QUESTION 6. If $f(x) = 2x - 3$ and $g(x) = x^2 + x$, then $f(g(x))$ equals

A) $2x^3 - x^2 - 3x$

B) $x^2 + 3x - 3$

C) $2x^2 + 2x - 3$

D) $4x^2 - 10x + 6$

E) None of the above.

COMPUTER ANALYSIS:

```
ITEM NUMBER    6              TEST NUMBER
FREQUENCY DISTRIBUTION
 GROUP          CHOICES
           0  A = 1  B = 2  C = 3  D = 4  E = 5
    I---------------------------------------------I
 1  I    2    19     1     1     3     3   I
 2  I    0    12     0     2     0     2   I
 3  I    1    11     2     9     0     1   I
 4  I    0     5     0     4     1     2   I
 5  I    0     3     0    26     1     0   I
    I---------------------------------------------I
    CORRECT RESPONSE IS   3.
```

```
 1  X **                         +
 2  X ******                     +
 3  X *******************        +
 4  X ****************** +
 5  X***************************************************
    ----+----+----+----+----+----+----+----+----+----+
PCT    10   20   30   40   50   60   70   80   90  100
```

```
ITEM-TEST R(CORRCTD)  0.556
DIFFICULTY      37.84
```

partitions:

Part I.   Questions 1 through 22.

Part II.   Questions 23 through 46.

Part III.   Questions 47 through 63.

The scores were interpreted as follows:

1.   Students who scored 18 or more correct on Part I were placed out of the first level (intermediate algebra) and into the second level.

2.   Students who scored 34 or better correct from Parts I and II were placed out of the second level and into the third level.

3.   Students who scored 11 or better correct on Part III and 45 or better correct on the total test were placed into calculus.

Evaluation of Exam.   To obtain statistical evaluation of the test's performance was a more complicated matter. Since SIU-C has no compulsory placement procedures, the students do not have to take a placement test and they may ignore the results of the test if they do take it. To obtain significant data, records were searched for students who took the course

248 MATHEMATICS EDUCATION

recommended to them by their performance on the placement test. For these students, scores on the placement test compared with their final grades for the course taken is shown in Table 2.

Table 2.

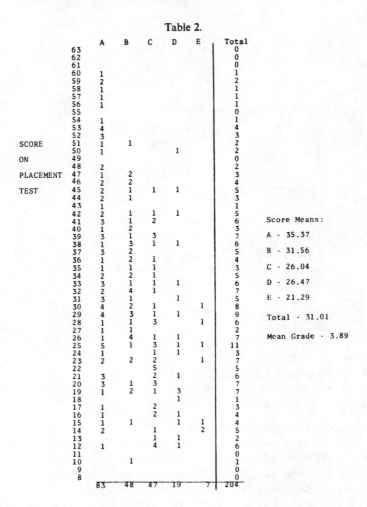

		A	B	C	D	E	Total
	63						0
	62						0
	61						0
	60	1					1
	59	2					2
	58	1					1
	57	1					1
	56	1					1
	55						0
	54	1					1
	53	4					4
	52	3					3
SCORE	51	1	1				2
	50	1			1		2
ON	49						0
	48	2					2
PLACEMENT	47	1	2				3
	46	2	2				4
TEST	45	2	1	1	1		5
	44	2	1				3
	43	1					1
	42	2	1	1	1		5
	41	3	1	2			6
	40	1	2				3
	39	3	1	3			7
	38	1	3	1	1		6
	37	3	2				5
	36	1	2	1			4
	35	1	1	1			3
	34	2	2	1			5
	33	3	1	1	1		6
	32	2	4	1			7
	31	3	1		1		5
	30	4	2	1		1	8
	29	4	3	1	1		9
	28	1	1	3		1	6
	27	1	1				2
	26	1	4	1	1		7
	25	5	1	3	1	1	11
	24	1		1	1		3
	23	2	2	2		1	7
	22			5			5
	21	3		2	1		6
	20	3	1	3			7
	19	1	2	1	3		7
	18				1		1
	17	1		2			3
	16	1		2	1		4
	15	1	1		1	1	4
	14	2		1		2	5
	13			1	1		2
	12	1		4	1		6
	11						0
	10		1				1
	9						0
	8						0
		83	48	47	19	7	204

Score Means:

A - 35.37

B - 31.56

C - 26.04

D - 26.47

E - 21.29

Total - 31.01

Mean Grade - 3.89

As the table indicates 204 students were located. 178 students or 87% of the total received a C grade or better in their course taken. And over 40% received an A grade. This may be considered significant since the usual percentage of students receiving a C grade or better in these courses is between 55% and 65%.

Table 3 indicates that the placement test constructed can also provide information concerning the validity of scores obtained on certain national placement exams related to a specific institution. The data in Table 3 show

Table 3.

STUDENT	PLACEMENT	PLACEMENT SCORE	CLEP SCORE
1		29	604
2		30	597
3	College Algebra &	34	604
4	Trigonometry I	32	660
5		32	628
6		25	510
7		29	589
8		27	636
9		31	581
MEAN		30	601
1		9	597
2		19	533
3	Intermediate	15	510
4	Algebra	22	518
5		18	565
6		28	494
7		14	533
8		32	589
9		27	549
10		27	626
11		28	676
12		20	502
13		29	580
14		23	589
15		26	533
16		28	518
17		23	597
18		29	526
MEAN		23	558
1		45	683
2		40	620
3	Calculus	41	691
4		45	707
5		41	620
6		39	683
7		45	800
8		42	731
9		40	660
10		40	683
11		43	800
12		45	786
MEAN		43	705
1		35	612
2		35	589
3	College Algebra &	33	680
4	Trigonometry II	28	620
5		39	668
6		38	719
7		43	707
8		35	612
9		37	628
10		31	644
11		41	581
12		30	618
13		36	565
14		33	557
15		33	604
16		32	612
17		36	652
18		34	581
19		32	446
20		35	652
21		39	652
22		43	742
23		38	723
24		35	510
25		31	573
26		36	668
27		37	665
28		34	525
29		42	683
30		36	628
31		34	657
32		35	620
MEAN		36	625

that there is very little correlation between scores obtained on the CLEP exam* and scores obtained on the math department placement exam—especially in the areas of college algebra, trigonometry, and intermediate algebra. Since the departmental placement exam has shown itself to be reliable, a reevaluation of CLEP as a placement instrument on the SIU-C campus is in order. This will require gathering a larger amount of statistical data over a longer period of time.

Conclusions. Because of the lack of general agreement across campuses as to the content of similarly titled courses, standardized placement tests may not be meeting the needs of individual schools. The procedure of test construction and interpretation described here could be useful for developing mathematical placement exams tailored to individual campus needs.

For those interested, information on advanced placement exams in calculus is available from

> College Entrance Examination Board
> Educational Testing Service
> Princeton, New Jersey 08540

There is also a pamphlet titled "Making Your Own Test" published by the Cooperative Test Division of the Educational Testing Service, whose address is given above.

---

*CLEP is a college level entrance program examination published by the Princeton Educational Testing Service.

8. RECREATIONAL MATHEMATICS

> "Mathematics is a world created by the mind of man, and the mathematicians are people who devote their lives to a wonderful kind of play."
>
> —Constance Reid, Two-Yr. Coll. Math. J. 11(1980)

> "I don't know what I may seem to the world, but, as to myself, I seem to have been only as a boy playing on the seashore, and diverting myself in now and then finding a smoother pebble or a prettier shell than ordinary, whilst the great ocean of truth lay all undiscovered before me."
>
> —Isaac Newton quoted by Rev. J. Spence: Anecdotes, Observations, and Characters of Books and Men (London, 1858), 40

Mathematical problems, puzzles, and games have been popular since antiquity, and their solutions have contributed much to the development of modern mathematics. Thus, Leibniz appears to have been correct when he said "Men are never so ingenious as when they are inventing games." The theme of recreational mathematics, in fact, could serve as an effective educational vehicle for exploring the historical and evolutionary aspects of many important branches of mathematics.

Recreational mathematics is characterized by a sense of relaxed speculation and free functioning within prescribed constraints. This "play," so to speak, permeates and lies at the heart of all mathematics. The articles in this chapter illustrate the essence and pervasive role of recreational mathematics in terms of its problem-solving nature. Loosely speaking, a problem may be defined as an intricate, unsettled question raised for inquiry or solution.

Properly posed questions play an important role in mathematics education. The pedagogical technique of Guided Discovery, for example, uses questions to stimulate students to solve problems proportionate to their knowledge. For another example, consider the question of how to pictorially represent some basic notions in group theory. In his article "Aspects of

Group Theory in Residue Designs," Thomas Moore describes how to create visually interesting art forms that can be used to illustrate and study important properties of finite groups.

Richard Johnsonbaugh's article "de Bruijn's Packing Problem" nicely corroborates Pólya's conviction "A great discovery solves a great problem, but there is a grain of discovery in the solution of any problem." Johnsonbaugh's article describes how the Dutch mathematician N. G. de Bruijn solved his young son's problem of filling a $6 \times 6 \times 6$ box with $1 \times 2 \times 4$ bricks and, in the process, answered two related questions—all in arbitrary dimension.

One need only consider the widespread popularity of the problem sections in math journals in order to be convinced that recreational problem-solving stimulates the intellect and gives impetus to creative mathematical activities. Warren Page's article "A General Approach to $p \cdot q$ r-cycles" is a good case in point. A natural number N is called a $p.q$ r-cycle if the number formed by placing the last r digits first is the rational multiple p/q of N. Specific problems asking for smallest $p.q$ r-cycles continue to appear. in the mathematics literature. In his paper, Page introduces a general theory that extends the results of other authors and actually produces all $p.q$ r-cycles for $r = 1, 2, 3$.

A puzzle may be defined as a problem or contrivance designed for testing ingenuity. The next two articles illustrate this aspect of recreational mathematics. In "Coin Weighings Revisited," A. Buchman and F. Hawthorne use a modern result to solve a well-known folk problem that was popular in the mid 1940's. Their solution is strikingly unusual since it actually produces an algorithm for locating and identifying the nature of a counterfeit coin. The puzzle *Instant Insanity* challenges one to arrange four $1 \times 1 \times 1$ cubes into a $1 \times 1 \times 4$ stack so that each of four colors appears on each of the four vertical sides of the stack. In "How to Tree Instant Insanity," Jean Pedersen uses tree diagrams to deduce the solution to this and other similar puzzles.

A game is essentially a competitive puzzle or situation conducted according to specified rules. As is well known, games of chance led to the development of probability theory, and games of strategy (where each participant, and not chance alone, can influence the outcome) play an integral part of decision-making in business, the military, and the social and behavioral sciences. Most teaching techniques can be viewed as strategic games in learning. Thus, it seems appropriate to conclude this chapter, as it began, with a pedagogically strategic game. "The Game Beware," by F. Pedersen and K. Pedersen, can be used to illustrate basic mathematical concepts such as topology, continuity, path, and tree. The concluding section of their article offers suggestions for using this game to further explore these and other related mathematical concepts.

Since the level of this book precludes the inclusion of highly esoteric articles which reflect the "play-like" nature of recreational mathematics, it may be of interest to quote A. Clebsch [*On Mathematics and Mathematicians*, Dover (1942), 99–100]:

> Research may start from definite problems whose importance it recognizes and whose solution is sought more or less directly by all forces. But equally legitimate is the other method of research which only selects the field of its activity and, contrary to the first method, freely reconnoiters in the search for problems which are capable of solution. If the first method leads to greater penetration, it is also easily exposed to the danger of unproductivity. To the second method we owe the acquisition of large and new fields, in which the details of many things remain to be determined and explored by the first method.

Aspects of Group Theory in Residue Designs

Thomas E. Moore, *Bridgewater State College, Bridgewater, MA*

Phil Locke [1] created a particularly intriguing form of visual art, called a residue design, by means of modular arithmetic. This art form can also be used to illustrate several important aspects of a finite group—as, for example, the coset decomposition of the group by one of its subgroups. We begin with a few examples.

For an odd prime p, let $Z_p^* = \{1, 2, \ldots, p-1\}$ denote the multiplicative group of residues modulo p. Locke's idea, which we apply to Z_p^*, distributes the elements of Z_p^*, equally spaced, in their natural (clockwise) order around a circle. Then one chooses a nonidentity multiplier $m \in Z_p^*$ and obtains all products $\{xm \pmod{p} : x \in Z_p^*\}$. Each pair of successive points, $x \in Z_p^*$ and $xm \pmod{p} \in Z_p^*$, on the circle are joined by a chord. Figure 1 shows this scheme for Z_{17}^* with $m = 2$.

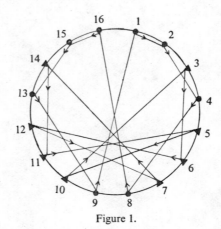

Figure 1.

Taking $x = 1$, for example, we follow the chordal path shown and obtain a "cyclic" polygon with vertices $\{1, 2, 4, 8, 16, 15, 13, 9\}$. This set H of vertices constitutes a cyclic subgroup of Z_{17}^* generated by $m = 2$. The other set of vertices, $\{3, 6, 12, 7, 14, 11, 5, 10\}$ corresponding to the cyclic polygon beginning with $x = 3$, is not a subgroup of Z_{17}^* (since it does not contain an identity element). This set, however, is the coset $3H$ of $H \subset Z_{17}^*$.

In general, one obtains a set of cyclic polygons which represent the cosets of a cyclic subgroup of Z_p^*. Figure 2 illustrates Z_{61}^* with $m = 3$. As above, one can exhibit the cyclic subgroup generated by $3 \in Z_{61}^*$ and show its 6 distinct cosets in Z_{61}^*.

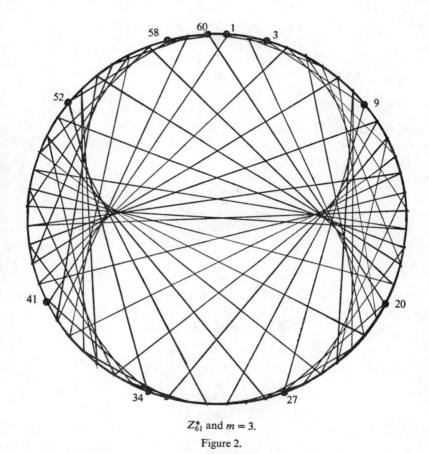

Z_{61}^* and $m = 3$.

Figure 2.

Although Locke's original creation concerned only modular arithmetic, we can consider similar designs using the elements of any finite group. For example, let S_n be the group of permutations of the symbols $\{1, 2, \ldots, n\}$ under composition of mappings, and let A_n denote the subgroup of even permutations in S_n. The two-rowed notation is standard for describing permutations—always from left to right. Thus,

$$\begin{pmatrix} 1 & 2 & 3 & 4 \\ 3 & . & . & . \end{pmatrix}\begin{pmatrix} 1 & 2 & 3 & 4 \\ . & . & 4 & . \end{pmatrix} = \begin{pmatrix} 1 & 2 & 3 & 4 \\ 4 & . & . & . \end{pmatrix}$$

occurs as part of

$$\begin{pmatrix} 1 & 2 & 3 & 4 \\ 3 & 2 & 1 & 4 \end{pmatrix}\begin{pmatrix} 1 & 2 & 3 & 4 \\ 2 & 3 & 4 & 1 \end{pmatrix} = \begin{pmatrix} 1 & 2 & 3 & 4 \\ 4 & 3 & 2 & 1 \end{pmatrix}.$$

For our design, consider the nonabelian group A_4 whose elements are

$$\iota = \begin{pmatrix} 1 & 2 & 3 & 4 \\ 1 & 2 & 3 & 4 \end{pmatrix} \quad \delta = \begin{pmatrix} 1 & 2 & 3 & 4 \\ 1 & 4 & 2 & 3 \end{pmatrix} \quad \rho = \begin{pmatrix} 1 & 2 & 3 & 4 \\ 4 & 1 & 3 & 2 \end{pmatrix}$$

$$\alpha = \begin{pmatrix} 1 & 2 & 3 & 4 \\ 2 & 1 & 4 & 3 \end{pmatrix} \quad \epsilon = \begin{pmatrix} 1 & 2 & 3 & 4 \\ 1 & 3 & 4 & 2 \end{pmatrix} \quad \vartheta = \begin{pmatrix} 1 & 2 & 3 & 4 \\ 2 & 4 & 3 & 1 \end{pmatrix}$$

$$\beta = \begin{pmatrix} 1 & 2 & 3 & 4 \\ 3 & 4 & 1 & 2 \end{pmatrix} \quad \lambda = \begin{pmatrix} 1 & 2 & 3 & 4 \\ 3 & 2 & 4 & 1 \end{pmatrix} \quad \sigma = \begin{pmatrix} 1 & 2 & 3 & 4 \\ 2 & 3 & 1 & 4 \end{pmatrix}$$

$$\gamma = \begin{pmatrix} 1 & 2 & 3 & 4 \\ 4 & 3 & 2 & 1 \end{pmatrix} \quad \phi = \begin{pmatrix} 1 & 2 & 3 & 4 \\ 4 & 2 & 1 & 3 \end{pmatrix} \quad \tau = \begin{pmatrix} 1 & 2 & 3 & 4 \\ 3 & 1 & 2 & 4 \end{pmatrix}.$$

Choosing δ as the left-multiplier, we create the design in Figure 3 which isolates the subgroup $H = \{\iota, \delta, \epsilon\}$ and its three other distinct left cosets $H\alpha, H\beta,$ and $H\gamma$ in A_4.

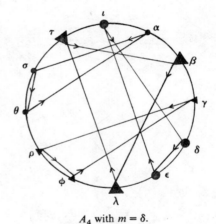

A_4 with $m = \delta$.

Figure 3.

Note. When δ is the right-multiplier, we obtain $\{\iota, \delta, \epsilon\}$ and its three distinct right cosets $\{\alpha, \rho, \lambda\}$, $\{\beta, \sigma, \phi\}$, and $\{\gamma, \lambda, \theta\}$ in A_4.

The group-theoretic content of a residue design is easily explained. Let G be a finite group with identity e, whose elements are equally spaced around a circle. For any nonidentity multiplier $m \in G$, the chordal path joining successive endpoints $\{e, em = m, mm = m^2, m^2m = m^3, \ldots, m^k = e\}$ is seen to generate a closed cyclic polygon whose vertices $H = \{e, m, m^2, \ldots, m^{k-1}\}$ constitute a kth order cyclic subgroup of G generated by m. For any element $x \in G - H$, the coset $xH = \{x, xm, xm^2, \ldots, xm^{k-1}\}$

of H clearly appears as the set of vertices of some cyclic polygon in our design. In fact, every such coset of H is realized as a polygon in the residue design. Accordingly, the number of cyclic polygons in a design is the index of the subgroup generated by the multiplier. All cyclic polygons have the same number of vertices—namely, the order of the multiplier in this group.

Remark. An alternate technique for obtaining designs using modular arithmetic is considered in [2].

REFERENCES

1. Phil Locke, Residue designs, Math. Teacher, 65 (1972), 260–263.
2. Daniel Shanks, Solved and Unsolved Problems in Number Theory, Spartan Books, Washington, D.C., 1962.

De Bruijn's Packing Problem

Richard Johnsonbaugh, *Chicago State University Chicago, IL*

In 1969, the Dutch mathematician N. G. de Bruijn proved three theorems (see [1]) which he said were inspired by his seven-year-old son who could not fill his $6 \times 6 \times 6$ box with $1 \times 2 \times 4$ bricks. De Bruijn's results solved his son's box-packing problem as well as answered two related questions— all in arbitrary dimensions. Here we prove de Bruijn's three theorems in two dimensions, the first of which describes when an $a \times b$ rectangle can be covered by $k \times m$ rectangles. Interested readers may wish to attempt the generalized proofs for n-dimensions, or they may find them in the above reference.

A covering of rectangle R by congruent rectangles R^* shall mean that R consists of nonoverlapping rectangles R^* where edges are parallel to the edges of R. Figure 1, for instance, shows a 7×12 rectangle covered by 3×4 rectangles.

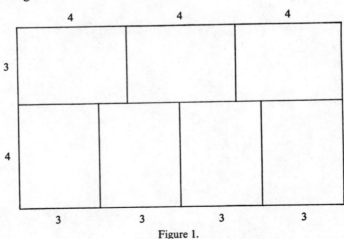

Figure 1.

The key idea behind de Bruijn's proofs is the generalization of a solution to the well-known problem:

Remove two diagonally opposite squares from an 8×8 checkerboard, and show that the remaining 62 squares cannot be covered by 31 dominoes of size 1×2.

The solution involved uses the observation that each domino covers exactly one black and one white square, and that these 31 dominoes are required to cover 30 squares of one color and 32 squares of the other color.

We shall call a collection S of integers k-balanced (or just balanced if the integer k is understood) if S consists of the integers $\{1, 2, \ldots, k\}$, each repeated the same number of times. Note that the set difference $S - T$, of k-balanced sets S and T, is also k-balanced.

The following lemma lies at the heart of de Bruijn's theorems. Here, and throughout this paper, | means "divides."

Lemma. *If an $a \times b$ rectangle is covered by $1 \times k$ rectangles, then $k | a$ or $k | b$.*

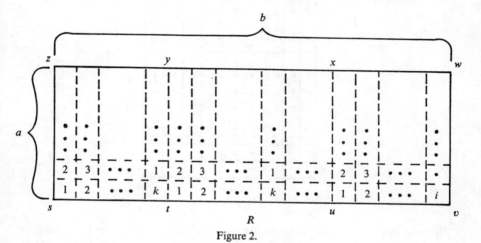

Figure 2.

Proof. Suppose, by way of contradiction, that $k \nmid a$ and $k \nmid b$. Partition R into $a \cdot b$ unit squares, and number them (mod k) from left to right and from bottom to top as in Figure 2. Since R is covered by $1 \times k$ rectangles, each rectangle covers each of the numbers $1, 2, \ldots, k$ exactly once. Therefore, the collection of integers in R is balanced. Each row of rectangle $R^* = styz$ (Figure 2) contains each of the numbers $1, 2, \ldots, k$ once, and so R^* is balanced. Therefore, $R - R^*$ is balanced. Now eliminate all such rectangles R^* with bottom row labeled $1, 2, \ldots, k$. If we eliminate the entire rectangle R, then $i = k$ and $k | b$. Therefore, by assumption $i < k$. Now rectangle $uvpo$ of Figure 3 is balanced, and we may remove it from the balanced rectangle $uvwx$ to obtain a balanced rectangle $opwx$. Similarly, we remove all rectangles from $uvwx$ whose left-hand column reads $1, 2, \ldots, k$. If we eliminate the whole rectangle $uvwx$, then $j = k$ and $k | a$. Accordingly, by assumption, $j < k$. Since rectangle $rqwx$ is balanced, each integer $1, 2, \ldots, k$ is repeated the same number of times, say m. Therefore, the $i \cdot j$ boxes contain $m \cdot k$ integers—one in each box. In other words,

$$i \cdot j = m \cdot k \quad \text{for} \quad 1 \leqslant m \leqslant k - 1.$$

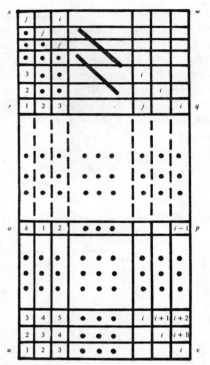

Figure 3.

But the integer i occurs only on its diagonal (since $(i \pm h) - k = i$ implies that $h = \pm k$, and this contradicts the fact that there are only i boxes in each row of $rqwx$). And, the integer k occurs only on its diagonal. Since i and j occur the same number $\min\{i, j\}$ of times in $rqwx$, we have $m = \min\{i, j\}$ and either $i = k$ or $j = k$. This contradiction establishes that either $k|a$ or $k|b$ as asserted.

It is now a simple matter to prove de Bruijn's first result.

Theorem 1. *If an $a \times b$ rectangle is covered by $n \times k$ rectangles, then $k \,|\, a$ or $k \,|\, b$, and $n \,|\, a$ or $n \,|\, b$.*

Proof. Suppose the $a \times b$ rectangle R is covered by $n \times k$ rectangles. Each $n \times k$ rectangle may be subdivided into n rectangles of size $1 \times k$ as well as into k rectangles of size $1 \times n$. Hence, R is covered by $1 \times k$ rectangles as well as by $1 \times n$ rectangles.

The three-dimensional version of Theorem 1 assured de Bruijn that his son could not fill his $6 \times 6 \times 6$ box with $1 \times 2 \times 4$ bricks since $4 \nmid 6$. De

Bruijn's next result uses the following terminology[†]: An $a \times b$ rectangle is a multiple of a $k \times m$ rectangle if either $k \mid a$ and $m \mid b$, or $k \mid b$ and $m \mid a$. (Any rectangle R which is a multiple of a rectangle R^* is easily seen to be covered by rectangles R^*.) We call a $k \times m$ rectangle harmonic if $k \mid m$ or $m \mid k$.

Theorem 2. *If a rectangle R can be covered by harmonic rectangles R^*, then R is a multiple of R^*.*

Proof. Suppose R is $a \times b$, and let R^* be $k \times m$ with $k \mid m$. By Theorem 1, either $m \mid a$ or $m \mid b$. Assume that $m \mid a$. To see that $k \mid b$, simply note that the edge of length b of R is a combination of the length k and the length a corresponding to some horizontal and vertical arrangement of the $k \times m$ covering rectangles. Since m is a multiple of k, it follows that b is also a multiple of k.

The three-dimensional analog of Theorem 2 asserts that the only boxes which can be filled by $i \times j \times k$ harmonic bricks are those of dimension $ai \times bj \times ck$. Thus, $1 \times 2 \times 4$ bricks can only fill out boxes of dimension $a \times 2b \times 4c$.

We conclude with the following fact about nonharmonic rectangles.

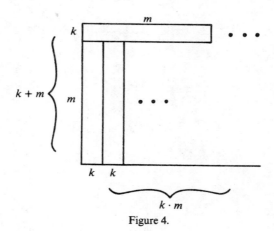

Figure 4.

[†]This terminology generalizes to n-dimensions. For example, the $1 \times 2 \times 4$ brick referred to earlier is a harmonic three-dimensional brick. (And $2 \times 1 \times 6 \times 6$ is a harmonic four-dimensional brick.) The three-dimensional analog of an $a \times b \times c$ brick R being a multiple of an $i \times j \times k$ brick R^* is that one of the lengths of R^* (say, j) divides one of the lengths of R (say, c) and the remaining two-dimensional $a \times b$ rectangle of R is a multiple of the remaining $i \times k$ rectangle of R^*. Thus, for example, $10 \times 16 \times 17$ is a multiple of $1 \times 2 \times 4$.

Theorem 3. *For each nonharmonic rectangle* R^*, *there is a rectangle* R *which is not a multiple of* R^*, *but which can be covered by rectangles* R^*.

Proof. Suppose R^* is $k \times m$, where $k < m$ and $k \nmid m$. The $(k + m) \times (k \cdot m)$ rectangle R is not a multiple of R^* (since neither k nor m divides $k + m$), yet R can be covered by rectangles R^* as shown in Figure 4.

Remark. Note that a 4×12 rectangle is both covered by, and a multiple of, the nonharmonic 2×3 rectangles.

REFERENCE
1. N. G. de Bruijn, Filling boxes with bricks, Amer. Math. Monthly, 76 (1969) 37-40.

A General Approach to $p.q$ r-cycles

Warren Page, *New York City Technical College, Brooklyn, NY*

In this article, all letters will denote nonnegative integers. It will be convenient to abbreviate k-digit numbers

$$N = n_1 \cdot 10^{k-1} + n_2 \cdot 10^{k-2} + \cdots + n_k$$

as $N = n_1 n_2 \ldots n_k$.

Suppose that you are asked to find the number $N = n_1 n_2 \ldots n_k$, with the fewest digits, such that the number formed by placing the last digit first is $\frac{6}{5}$ of N. Was your solution to

$$5(n_k n_1 n_2 \ldots n_{k-1}) = 6(n_1 n_2 \ldots n_k)$$

the number $N = 45$? How about finding the number N, with the fewest digits, such that the number formed by placing the last two digits first is $\frac{3}{4}$ of N? Here, the desired answer to

$$4(n_{k-1} n_k n_1 n_2 \ldots n_{k-2}) = 3(n_1 n_2 \ldots n_k)$$

is $N = 108$.

Let us call a number $N = n_1 n_2 \ldots n_k$ a $p.q$ r-cycle (p, q are nonzero digits) whenever

$$p(n_{k-r+1} n_{k-r+2} \cdots n_k n_1 n_2 \ldots n_{k-r}) = q(n_1 n_2 \ldots n_k).$$

Thus, 45 is a 5.6 1-cycle and 108 is a 4.3 2-cycle.

Problems concerning $p.q$ r-cycles have always been popular in the mathematics literature, and there are good reasons for this. Since four parameters $\{p, q, r, k\}$ are involved, many interesting questions and conjectures arise naturally. The problem of Trigg [10], for example, yielded a 3.4 3-cycle 428571 when $k = 6$. And Hunter's quest [1] for the smallest 1.8 $(n + 1)$-cycle with $k = 2n + 1$ surprisingly turns out to be the 35-digit 9.8 1-cycle in [7]. Wlodarski's solution [11] for determining the smallest 1.6 1-cycle with $n_k = 6$ was extended by Klamkin [5] to characterize the smallest 1.6 1-cycles with arbitrary n_k. Other related results continue to appear in the literature (see [2], [3], [4], [6], [7], [8], for example).

Our intent is to draw together and extend the work of the above authors. Here we shall introduce a general approach to $p.q$ r-cycles and actually produce all $p.q$ r-cycles for $r = 1, 2, 3$.

1. General Results. Notationwise, $n_1 \cdot n_2$ means n_1 times n_2, whereas $n_1 n_2$ denotes the two-digit number $10 \cdot n_1 + n_2$. Multiplication and juxtaposition are respectively written $(n_1 n_2 \ldots n_\alpha) \cdot (m_1 m_2 \ldots m_\beta)$ and $\{n_1 n_2 \ldots n_\alpha\}\{m_1 m_2 \ldots m_\beta\}$. Thus, $\{135\}\{4 \cdot (26)\} = 135104$. Since N is a $p.q$ r-cycle if and only if it is a $(p/(p,q)).(q/(p,q))$ r-cycle [here (p,q) denotes the greatest common factor of p and q], it is henceforth assumed that $p \neq q$ satisfy $(p,q) = 1$. Finally, N is a $p.q$ r-cycle if and only if $N^i = \{N\}\{N\} \ldots \{N\}$ (i-times) is a $p.q$ r-cycle for every natural number i. Unless specified otherwise, we always assume that $N = N^1$.

By definition, $N = n_1 n_2 \ldots n_k$ is a $p.q$ r-cycle if and only if

$$p \cdot 10^{k-r} \cdot (n_{k-r+1} n_{k-r+2} \ldots n_k) + p \cdot (n_1 n_2 \ldots n_{k-r})$$
$$= q \cdot 10^r \cdot (n_1 n_2 \ldots n_{k-r}) + q \cdot (n_{k-r+1} n_{k-r+2} \ldots n_k).$$

This can be simplified as

$$n_1 n_2 \ldots n_{k-r} = (n_{k-r+1} n_{k-r+2} \ldots n_k) \cdot \left(\frac{p \cdot 10^{k-r} - q}{q \cdot 10^r - p} \right). \tag{1}$$

Similarly, $N = n_1 n_2 \ldots n_k$ is a $p.q$ r-cycle if and only if

$$N = \frac{p \cdot (n_{k-r+1} n_{k-r+2} \ldots n_k)}{q \cdot 10^r - p} \cdot (10^k - 1) \tag{2}$$

since (2) is equivalent to requiring that N satisfy

$$q \cdot N = p \cdot \left[\frac{N - (n_{k-r+1} n_{k-r+2} \ldots n_k)}{10^r} + 10^{k-r} \cdot (n_{k-r+1} n_{k-r+2} \ldots n_k) \right].$$

It is clear, from $10^r \cdot (p \cdot 10^{k-r} - q) = p \cdot (10^k - 1) - (q \cdot 10^r - p)$, that

$$(q \cdot 10^r - p, 10^k - 1) | (p \cdot 10^{k-r} - q)$$

and

$$(q \cdot 10^r - p, p \cdot 10^{k-r} - q) | p \cdot (10^k - 1). \tag{3}$$

Our first result (extending Theorem 6 of [7], and Problem 29 of [8]) may be surprising.

Theorem 1. If $N = n_1 n_2 \ldots n_k$ is a $p.q$ r-cycle and $1 \leqslant r \leqslant 3$, there can be no $p.q$ r-cycle $M = m_1 m_2 \ldots m_{k+1}$.

Proof. Assume that M is a $p.q$ r-cycle. It follows from (2) that

$$\frac{p \cdot (n_{k-r+1} \ldots n_k)}{10^r - p} = \frac{N}{10^k - 1} < 1$$

and

$$\frac{p \cdot (m_{k-r+2} \ldots m_{k+1})}{10^r - p} = \frac{M}{10^{k+1} - 1} < 1.$$

Therefore, there exist prime factors x, y of $q \cdot 10^r - p$ such that $x | (10^k - 1)$ and $y | (10^{k+1} - 1)$. But $(10^k - 1, 10^{k+1} - 1) = 9$, and $9 \nmid (q \cdot 10^r - p)$ (since

$q \cdot 10^r - p = (q - p)(\bmod 9)$ and $q \neq p)$. Thus, one of the following must hold:

(i) $q \cdot 10^r - p$ contains distinct prime factors $x, y > 3$ satisfying

$$x \mid (10^k - 1) \quad \text{and} \quad y \mid (10^{k+1} - 1),$$

(ii) $q \cdot 10^r - p = 3 \cdot Z_k$ or $q \cdot 10^r - p = 3 \cdot Z_{k-1}$, where Z_t denotes a product of primes none of whose factors divide $10^t - 1$.

If $q \cdot 10^r - p = 3 \cdot Z_k$, then

$$N = \frac{p \cdot (n_{k-r+1} \cdots n_k)}{Z_k} \cdot \left(\frac{10^k - 1}{3} \right)$$

is a chain of k digits $aa \ldots a$ $(1 \leqslant a < 9)$ and we obtain the contradiction that $p = q$. Similarly, $q \cdot 10^r - p = 3 \cdot Z_{k+1}$ implies that

$$M = \frac{p \cdot (m_{k-r+2} \cdots m_{k+1})}{Z_{k+1}} \cdot \left(\frac{10^{k+1} - 3}{3} \right)$$

equals $bb \ldots b$ $(1 \leqslant b < 9)$ and $p = q$. The foregoing remarks show that only case (i) prevails. Since juxtaposition is omitted, we may assume that the prime factors $x, y > 3$ in (i) satisfy

$$k = \lambda(x) \quad \text{and} \quad k + 1 = \lambda(y)$$

where $\lambda(j)$ denotes the least exponent ϵ such that $j \mid (10^e - 1)$. Using a table of least exponents (see [9], for example), we show that $\lambda(y) = \lambda(x) + 1$ fails for $1 \leqslant r \leqslant 3$.

If $r = 1$, then $1 \leqslant q \cdot 10 - p \leqslant 89$ and $7 \leqslant x, y \leqslant 11$. This clearly contradicts $\lambda(y) = \lambda(x) + 1$ since $\lambda(7) = 6$ and $\lambda(11) = 2$.

If $r = 2$, then $91 \leqslant q \cdot 10^2 - p \leqslant 899$ and $7 \leqslant x, y \leqslant 127$. For convenience, let $*$ denote "contradiction" and let $\Rightarrow$ denote "implication."

$x = 7 \Rightarrow 6 = \lambda(x)$. Thus, $7 = \lambda(y) \Rightarrow y > 127 *$.

$y = 7 \Rightarrow 6 = \lambda(y)$, and so $5 = \lambda(x) \Rightarrow x = 41$ (or $x > 127 *$). Evidently $q \cdot 10^2 - p = \{q - 1\}\{9\}\{10 - p\}$ does not equal $41 \cdot 7 \cdot z$ for $z = 1, 2, 3$.

Now $11 \nmid (q \cdot 10^2 - p)$ since $q \cdot 10^2 - p = (q - p)(\bmod 11)$ and $p \neq q$. Therefore, $13 \leqslant x, y \leqslant 67$ and, since $\lambda(13) = 6$, the same reasoning reduces our range to $17 \leqslant x, y \leqslant 47$.

$x = 17 \Rightarrow 16 = \lambda(x)$. Thus, $17 = \lambda(y) \Rightarrow y > 47 *$.

$y = 17 \Rightarrow 15 = \lambda(x)$ and $x = 31$ (or $x > 47 *$). It follows, from $q \cdot 10^2 - p \neq 31 \cdot 17$, that $19 \leqslant x, y \leqslant 47$.

$x = 19 \Rightarrow 18 = \lambda(x)$, and $19 = \lambda(y) \Rightarrow y > 47 *$.

$y = 19 \Rightarrow 17 = \lambda(x)$ and $x > 47 *$. Thus, $23 \leqslant x, y \leqslant 37$.

$x = 23 \Rightarrow 22 = \lambda(x)$, and $23 = \lambda(y) \Rightarrow y > 37 *$.

$y = 23 \Rightarrow 21 = \lambda(x)$ and $x > 37 *$.

Finally, $29 \cdot 31 = 899$ and $|\lambda(29) - \lambda(31)| = |28 - 15| \neq 1$. There-
fore, all possibilities have been exhausted and M cannot be a $p.q$
2-cycle.

If $r = 3$, then $991 \leqslant q \cdot 10^3 - p \leqslant 8999$ and $7 \leqslant x, y \leqslant 1283$. The same
reasoning as above can be used to verify that $\lambda(y) \neq \lambda(x) + 1$ for all pairs
x, y in this range, Therefore, M cannot be a $p.q$ 3-cycle.

Our proof is now complete; it remains only to observe that both
$N = 023255813953488372093$ and $M = 043478260869565217313913$ are 1.9
4-cycles.

Remark. The counterexample for $r = 4$ was obtained by considering
$r = 2s$ and using $x = (\sqrt{q} \cdot 10^s + \sqrt{p}, 10^k - 1)$ and $y = (\sqrt{q} \cdot 10^s - \sqrt{p},$
$10^{k+1} - 1)$, or $x = (\sqrt{q} \cdot 10^s + \sqrt{p}, 10^k - 1)$ and $y = (\sqrt{q} \cdot 10^s - \sqrt{p}, 10^{k+1} -$
$1)$. Since $9 \cdot 10^4 - 1 = 301 \cdot 299 = (7 \cdot 43) \cdot (13 \cdot 23)$ and $\lambda(23) = 22 =$
$\lambda(43) + 1$, we used (2) to write out M and N. Similarly, $9 \cdot 10^6 - 1 = 3001 \cdot$
2999 and $\lambda(3001) = 1500 = \lambda(2999) + 1$ establishes the existence of 1499
and 1500-digit 9.1 6-cycles.

One way to generate $p.q$ r-cycles is based on the following:

Definition 2. A $p.q$ r-cycle N is an *mth order generator* $(2 \leqslant m \leqslant 9)$,
denoted $G_{p.q}^{(m)}$, if $j \cdot N$ is a $p.q$ r-cycle for $j = 2, 3, \ldots, m$. We term $G_{p.q}^{(9)}$ a
generator, and abbreviate it as $G_{p.q}$.

Remark. The property of being the smallest $p.q$ r-cycle is neither necessary nor sufficient for being a generator. For example, $G_{3.8} = 038961$ is
greater than the smallest 3.8 1-cycle $N = 27$, and N is not even a second
order generator.

Let N' denote $n_{k-r+1} n_{k-r+2} \cdots n_k n_1 \cdots n_{k-r}$ for a $p.q$ r-cycle $N =$
$n_1 n_2 \ldots n_k$. Theorem 1 enables us to completely characterize generators.

Theorem 3. *A $p.q$ r-cycle* $(1 \leqslant r \leqslant 3)$ $N = n_1 n_2 \ldots n_k$ *is a generator if
and only if it has one of the following forms:*

$$G_0 = 0 n_2 n_3 n_4 \ldots n_{k-r} \delta n_{k-r+2} \cdots n_k,$$
$$G_1 = 10 n_3 n_4 \ldots n_{k-r} 0 n_{k-r+2} \cdots n_k,$$
$$G_2 = 110 n_4 \ldots n_{k-r} 0 n_{k-r+2} \cdots n_k,$$

where either $\delta = 0$, *or else* $n_{k-r+1} n_{k-r+2} \cdots n_k$ *is a chain of ones, or*
$n_{k-r+1} n_{k-r+2} \cdots n_k$ *begins with a chain of ones followed by a zero.*

Proof. First note that N is an mth order generator if and only if $j \cdot N'$
$= (j \cdot N)'$ for $j = 2, 3 \ldots m$. To be sure,

$$\frac{j \cdot N}{j \cdot N'} = \frac{N}{N'} = \frac{p}{q}$$

since N is a $p.q$ r-cycle, and N is an mth order generator if and only if

$(p/q) = (j \cdot N)/(j \cdot N)'$ for all $j = 2, 3, \ldots, m$. Now $j \cdot N' = (j \cdot N)'$ for every natural number j satisfying $j \cdot (n_1 n_2 \ldots n_{k-r}) < 10^{k-r+1}$ and $j \cdot (n_{k-r+1} n_{k-r+2} \ldots n_k) < 10^{r+1}$ since these inequalities assure that

$$(j \cdot N)' = [j \cdot 10^r \cdot (n_1 n_2 \ldots n_{k-r}) + j \cdot (n_{k-r+1} n_{k-r+2} \ldots n_k)]'$$

$$= j \cdot 10^{k-r} \cdot (n_{k-r+1} n_{k-r+2} \ldots n_k) + j \cdot (n_1 n_2 \ldots n_{k-r}) = j \cdot N'.$$

The preceding remarks establish that the $G_{i>0}$ above are all generators. Conversely, suppose that N is a generator. Then N begins with zero or a chain of ones followed by a zero (otherwise the $k + 1$ digit number $9 \cdot N$ is not a *p.q* *r*-cycle). Furthermore, $n_1 = 0$ necessitates that $\delta = 0$ or $n_{k-r+1} n_{k-r+2} \ldots n_k$ have one of the forms specified above. If not, the $k + 1$ digit number $9 \cdot N'$ is strictly larger than the (at most k digit) number $(9 \cdot N)'$. Similarly, $n_{k-r+1} = 0$ if $n_1 = 1$ since otherwise, $p \cdot N' = p x_2 x_3 \ldots x_k$ cannot equal $q y_2 y_3 \ldots y_k = q \cdot N$. Finally, there is no generator of the form $G_3 = 1110 n_5 \ldots n_{k-r} 0 n_{k-r+2} n_k$. When $r = 3$, for example, $p \cdot G_3' = q \cdot G_3$ requires that $p \cdot (n_{k-1} n_k) = 3 \cdot 37 \cdot q$ and $p = 3$. Therefore, either $q = 1$ and $N_{3.1} = [3 \cdot (037)/(10^3 - 1)] \cdot (10^k - 1) = 1113 n_5 \ldots n_{165} 9$, or $q = 2$ and $N_{3.2} = [3 \cdot (074)/(2 \cdot 10^3 - 3)] \cdot (10^k - 1) = 11116 n_6 \ldots n_{997} 6$.

Remark. Since $p/q = G_i/G_i'$, it follows that $G_{i>0}$ exists only if $p > q$. And G_0, with $\delta = 1$, exists only if $p < q$.

2. p.q 1-cycles. All 1-cycle generators are of the form

$$G_0 = 0 n_2 n_3 \ldots n_{k-1} 1$$

since $\delta = n_k \neq 0$. One technique for producing *p.q* 1-cycles uses (1) and the fact (3) that $\lambda(x) \mid (p \cdot 10^{k-r} - q)$ for $x \mid (q \cdot 10^r - p)$.

Example. (i) The search for a 1.4 1-cycle begins with

$$n_1 n_2 \ldots n_{k-1} = 1 \cdot \left[\frac{1 \cdot 10^{k-1} - 4}{4 \cdot 10 - 1} \right]$$

Since $k = \lambda(39) = 6$, we obtain

$$G_{1.4} = \left\{ 1 \cdot \left[\frac{10^5 - 4}{39} \right] \right\} \{1\} = \{02564\}\{1\} = 025641.$$

(ii) There does not exist a 5.8 1-cycle since

$$n_1 n_2 \ldots n_{k-1} = n_k \cdot \left[\frac{5 \cdot 10^{k-1} - 8}{8 \cdot 10 - 5} \right]$$

and $5 \nmid (5 \cdot 10^{k-1} - 8)$.

Using this technique, one can easily verify that each element in the table below is the asserted *p.q* 1-cycle. Although $N_{1.5} = 142857$ is not an integral multiple of $G_{1.5}$, we see that $N_{1.5}$ and $7 \cdot G_{1.5} = N_{1.5}{}^7$ are the only 1.5 1-cycles having $n_k = 7$. Similarly, $N_{3.8} = 27$ is not an integral multiple of $G_{3.8}$

$= 038961$, yet $N_{3.8}$ and $7 \cdot G_{3.8} = \overline{N_{3.8}}^3$ are the only 3.8 1-cycles with $n_k = 7$. This holds in general: *If $N_{p.q} = n_1 n_2 \ldots n_k$ and $G_{p.q}$ are p.q 1-cycles, then $n_k \cdot G_{p.q} = \overline{N_{p.q}}^j$ for some positive integer j.* Indeed,

$$N_{p.q} = \frac{p \cdot n_k}{q \cdot 10 - p} \cdot (10^k - 1) \quad \text{and} \quad n_k \cdot G_{p.q} = \frac{n_k \cdot p \cdot 1}{q \cdot 10 - p} \cdot (10^{k+t} - 1)$$

necessitates that $x = (10^k - 1, 10^{k+t} - 1)$ for some factor x of $q \cdot 10 - p$. Therefore, $k + t = j \cdot k$ and $n_k \cdot G_{p.q} = \overline{N_{p.q}}^j$.

These remarks, in effect, establish the following:

Theorem 4. *All p.q 1-cycles are obtained from the table on page 269 by taking digit multiples of $G_{p.q}^{(m)}$ and $G_{p.q}$.*

$G_{1.2} = 052631578947368421$
$G_{1.3} = 0344827586206896551724137931$
$G_{1.5} = 0204081632653061224489795981367346938777551$
$G_{1.6} = 0169491525423728813559322033898305084745762711864406779661$
$G_{1.7} = 014492753623188405797 1$
$G_{1.8} = 0126582278481$
$G_{1.9} = 011235955056179775280898876404494382022471 91$

$G_{2.7}^{(2)} = 1176470588235294$

$G_{3.2}^{(3)} = 1764705882352941$
$G_{3.5} = 0638297872340425531914893617021276595744680851$
$N_{3.7} = 44776119402985074626865671641791$

$G_{4.5}^{(4)} = 1739130434782608695652$
$G_{4.9}^{(4)} = 093023255813953488372$

$N_{5.9} = 2941176470588235$

$G_{7.3}^{(3)} = 3043478260869565217391$
$G_{7.5}^{(6)} = 16279069767441860465 1$
$N_{7.6} = 132075471698$
$G_{7.9} = 08433734939759036144578313253012048192771$

$G_{8.7}^{(3)} = 258064516129032$

$G_{9.4}^{(3)} = 290352580645161$
$G_{9.7}^{(6)} = 147540983606557377049180327868853459016393442622950819672131$
$G_{9.8}^{(7)} = 12670 6563380281690140845070422535 21$

p \ q	1	2	3	4	5	6	7	8	9
1		$G_{1,2}$	$G_{1,3}$	$G_{1,4} = 025641$	$N_{1,5} = 142857$ $G_{1,5}$	$G_{1,6}$	$G_{1,7}$	$G_{1,8}$	$G_{1,9}$
2			$G_{2,3}^{(2)} = 285714$	$G_{1,2}$		$G_{1,3}$	$G_{2,7}^{(2)}$	$G_{1,4}$	$N_{2,9} = 18$
3	$G_{3,1}^{(2)} = 428571$	$G_{3,2}^{(3)}$		$G_{3,4} = 081$	$G_{3,5}$	$G_{1,2}$	$G_{3,7}$	$N_{3,8} = 27$ $C_{3,8} = 038961$	$G_{1,3}$
4			$G_{4,3}^{(3)} = 307692$		$G_{4,5}^{(4)}$	$G_{2,3}^{(2)}$	$N_{4,7} = 38$	$G_{1,2}$	$G_{4,9}^{(4)}$
5				$N_{5,4} = 714285$		$N_{5,6} = 45$	$N_{5,7} = 384615$		$N_{5,9}$
6		$G_{3,1}^{(2)}$		$G_{3,2}^{(3)}$	$N_{6,5} = 54$			$G_{3,4}$	$G_{2,3}^{(2)}$
7		$N_{7,2} = 538461$	$G_{7,3}^{(3)}$	$G_{7,4}^{(4)} = 21$	$G_{7,5}^{(6)}$	$N_{7,6}$		$G_{7,8} = 09589041$	$G_{7,9}$
8			$N_{8,3} = 72$			$G_{4,3}^{(3)}$	$G_{8,7}^{(3)}$		$G_{8,9}^{(4)} = 19512$
9		$N_{9,2} = 18$	$G_{3,1}^{(2)}$	$G_{9,4}^{(3)}$	$G_{9,5}^{(4)} = 21952$	$G_{3,2}^{(3)}$	$G_{9,7}^{(6)}$	$G_{9,8}^{(7)}$	

3. $p.q$ 2-cycles. All 2-cycle generators are of the form $G_0 = 0n_2 n_3 \ldots n_{k-2} \delta n_k$ or $G_1 = 10n_3 \ldots n_{k-2} 0 n_k$. There are no 2-cycle generators $G_2 = 110 n_4 \ldots n_{k-2} 0 n_k$ since

$$p \cdot (0 n_k 110 n_4 \ldots n_{k-2}) = p \cdot G_2' = q \cdot G_2 = q \cdot (110 n_4 \ldots n_{k-2} 0 n_k)$$

requires that $p \cdot n_k = 11 \cdot q$, and this is impossible. If $q \cdot 10^2 - p = (n_{k-1} n_k) \cdot x$, then expressions (1) and (3) yield

$$N_{p.q} = \left\{ (n_{k-1} n_k) \cdot \left(\frac{p \cdot 10^{\lambda(x)-2} - q}{q \cdot 10^2 - p} \right) \right\} \{ n_{k-1} n_k \}. \tag{4}$$

We shall use (4) to determine all 2-cycle generators.

(a) Let us first consider "00-type" generators $G_0 = 0 n_2 n_3 \ldots n_{k-2} 0 n_k$. For such generators, $p \cdot G_0' = q \cdot G_0$ clearly requires that

$$p \cdot n_k = q \cdot n_2 + R \leqslant 9 \cdot q + (q-1) = 10 \cdot q - 1. \tag{5}$$

Suppose $q = 1$.

If $p = 2$, then

$$G_{2.1} = \left\{ (0 n_k) \cdot \left(\frac{2 \cdot 10^{k-2} - 1}{10^2 - 2} \right) \right\} \{ 0 n_k \}.$$

Here $n_k = 2$ (therefore $n_2 = 4$ by (5)) and $\lambda(49) = 42$ yields

$$G_{2.1} = \left\{ \frac{2 \cdot 10^{40} - 1}{49} \right\} \{ 02 \}$$

$$= 0408163265306122448938775510204081632 65102.$$

Similarly,

$$G_{3.1} = \left\{ (01) \cdot \left(\frac{3 \cdot 10^{94} - 1}{97} \right) \right\} \{ 01 \} = 03092 n_6 \ldots n_{93} 701.$$

There is no 00-type generator for $p = 4$ since (5) requires that $n_k = 2$, and $2 \nmid (4 \cdot 10^{k-2} - 1)$. For $p \geqslant 5$, we have $n_k = 1$. Accordingly, there are no generators for $p = 5, 6, 8$ (since $5 \nmid (5 \cdot 10^{k-2} - 1)$, and n_k is even when p is even). This leaves only

$$G_{7.1} = 075268817204301 \quad \text{and} \quad G_{9.1} = 098901.$$

Suppose $q = 2$.

If $p = 1$, then

$$G_{1.2} = \left\{ (01) \cdot \left(\frac{10^{97} - 2}{199} \right) \right\} \{ 01 \} = 00502 n_6 \ldots n_{96} 201.$$

(Note that $n_k = 1$ in (5) yields $n_2 = 0$.) There are no 00-type generators for $p = 5$ since $n_k \neq 5$ by (5), and $5 \nmid (5 \cdot 10^{k-2} - 2)$. The remaining values of p produce

$$G_{7.2} = 032694 n_8 \ldots n_{189} 601 \quad \text{and} \quad G_{9.2} = 0471204 n_8 \ldots n_{92} 801.$$

Proceeding in this manner, one can easily obtain the remaining 00-type generators.

$$G_{1.3} = 00334448n_9 \ldots n_{63}301$$
$$G_{2.3} = 0134228n_8 \ldots n_{145}302$$
$$G_{5.3} = 0847457n_8 \ldots n_{55}305$$
$$G_{7.3} = 0238907n_8 \ldots n_{144}901$$

$$G_{1.4} = 002506265664160401$$
$$G_{3.4} = 0075566n_8 \ldots n_{96}801$$
$$G_{5.4} = 0632911392405$$
$$G_{7.4} = 01781178n_9 \ldots n_{43}201$$
$$G_{9.4} = 0230179n_8 \ldots n_{173}601$$

$$G_{1.5} = 00200400n_9 \ldots n_{495}501$$
$$G_{2.5} = 0080321n_8 \ldots n_{38}502$$
$$G_{3.5} = 0060362n_8 \ldots n_{207}501$$
$$H_{3.5} = 0422538n_8 \ldots n_{32}507$$
$$G_{6.5} = 024291497975708502$$
$$G_{7.5} = 0014197n_9 \ldots n_{109}501$$
$$G_{8.5} = 06504$$

$$G_{1.6} = 00166944n_9 \ldots n_{296}601$$
$$G_{5.6} = 0420168n_8 \ldots n_{45}605$$
$$G_{7.6} = 0118043n_8 \ldots n_{589}801$$

$$G_{1.7} = 00143061n_9 \ldots n_{229}701$$
$$G_{2.7} = 0057306n_8 \ldots n_{113}702$$
$$G_{3.7} = 0043041n_8 \ldots n_{77}901$$
$$G_{4.7} = 0459770n_8 \ldots n_{25}608$$
$$G_{5.7} = 0359712n_8 \ldots n_{43}705$$
$$G_{6.7} = 0172910n_8 \ldots n_{170}602$$
$$G_{8.7} = 0462427n_8 \ldots n_{40}104$$
$$G_{9.7} = 0130246n_8 \ldots n_{227}301$$

$$G_{1.8} = 00125156n_9 \ldots n_{365}801$$
$$G_{3.8} = 003764n_7 \ldots n_{196}601$$
$$G_{5.8} = 0335570n_8 \ldots n_{145}805$$
$$G_{7.8} = 0088272n_8 \ldots n_{57}401$$
$$G_{9.8} = 0113780n_8 \ldots n_{333}201$$
$$H_{9.8} = 0796460n_8 \ldots n_{109}407$$

$$G_{1.9} = 00111234n_9 \ldots n_{417}901$$
$$G_{2.9} = 0044543n_8 \ldots n_{29}902$$
$$G_{5.9} = 0279329n_8 \ldots n_{175}905$$
$$G_{7.9} = 0078387n_8 \ldots n_{411}701$$
$$G_{8.9} = 0358744n_8 \ldots n_{219}704$$

For $p.q = 1.4$, one can use $n_k = 7$ (therefore $n_2 = 1$) and obtain $H_{1.4}$ $= 017543859649122807$, which is $7 \cdot G_{1.4}$. It is also clear that $7 \cdot G_{3.5} = \overline{H_{3.6}}^6$ and $7 \cdot G_{9.8} = \overline{H_{9.8}}^3$. The general case follows as for 1-cycles: *If $N_{p.q} = n_1 n_1 \ldots n_{k-2} m_1 m_2$ and $G_{p.q}$ are $p.q$ 1-cycles, then $(m_1 m_2) \cdot G_{p.q} = \overline{N_{p.q}}^j$ for some positive integer j.*

(b) Let us now consider "01-type" generators $G_0 = 0n_2 n_3 \ldots n_{k-2}11$. For such generators, $p/q = (G_0)/(G_0') < 1$. Furthermore, $11 \mathbin{\big/} (q \cdot 10^2 - p)$ and so $q \cdot 10^2 - p$ divides $p \cdot 10^{k-2} - q$. Accordingly, there is no 01-type generator when $p = 5$ or when $p.q = $ even.odd. From (4), we now obtain

$$G_{1.2} = 05527n_6 \ldots n_{96}211 \qquad G_{1.7} = 015736n_7 \ldots n_{229}711$$
$$G_{1.3} = 03678n_6 \ldots n_{63}311 \qquad G_{3.7} = 04734n_6 \ldots n_{77}611$$
$$G_{1.4} = 027568922305764411 \qquad G_{1.8} = 01376n_6 \ldots n_{365}811$$
$$G_{3.4} = 08312n_6 \ldots n_{96}811 \qquad G_{1.8} = 04140n_6 \ldots n_{196}611$$
$$G_{1.5} = 022044n_7 \ldots n_{495}511 \qquad G_{7.8} = 097099621n_{10} \ldots n_{57}411$$
$$G_{3.5} = 06639n_6 \ldots n_{207}511 \qquad G_{1.9} = 012235n_7 \ldots n_{417}911$$
$$G_{1.6} = 018363n_7 \ldots n_{296}611 \qquad G_{7.9} = 08622n_6 \ldots n_{411}711$$

Let $G_{(p.q)_{00}}$ and $G_{(p.q)_{01}}$ denote 00-type and 01-type generators, respectively. One can easily verify that each $G_{(p.q)_{01}}$ above satisfies $G_{(p.q)_{01}} = 11 \cdot G_{(p.q)_{00}}$. Therefore, $G_{(p.q)_{00}}$ is an extended generator in the sense that $(m_1 m_2) \cdot G_{(p.q)_{00}}$ is a $p.q$ 2-cycle for every positive integer $m_1 m_2 < 10^2$ having $m_2 \neq 0$. An extended generator must have $n_2 = 0$ (Theorem 1). The converse, of course, is false. For example, $11 \cdot G_{7.5} = 015616n_7 \ldots n_{109}511$ is not even a 7.5 2-cycle since $11 < 7$.

(c) The only 2-cycle generators of the form $G_1 = 10n_3 \ldots n_{k-2}0n_k$ are

$$G_{4.3} = 108 \quad \text{and} \quad G_{8.3} = 1095804.$$

This is clear since $p/q = G_1/G_1' > 1$ and $p \cdot 10^{k-2} - q$ divided by $(q \cdot 10^2 - p)/(n_k, q \cdot 10^2 - p)$ must be of the form $10n_3 \ldots$. The only pairs $p.q$ satisfying this requirement are 4.3 and 8.3.

Since it suffices to take $N_{p.q} = \overline{N_{p.q}}^1$ when $(m_1 m_2) \cdot G_{p.q} = \overline{N_{p.q}}^j$ the remarks following (a) establish:

Theorem 5. *All $p.q$ 2-cycles are obtained from the 2-cycle generators listed above.*

4. $p.q$ 3-cycles and beyond.

Our technique for finding 1-cycles and 2-cycles extends to 3-cycles. To find Trigg's 3.4 3-cycle in [10], for example, consider

$$n_1 n_2 \ldots n_{k-3} = (n_{k-2} n_{k-1} n_k) \cdot \left(\frac{3 \cdot 10^{k-3} - 4}{4 \cdot 10^3 - 3} \right)$$

$$= (n_{k-2} n_{k-1} n_k) \cdot \left(\frac{3 \cdot 10^4 - 4}{7 \cdot 571} \right).$$

Since $\lambda(7) = 6$ and $\lambda(571) = 570$, there are two distinct possibilities: For $n_{k-2} n_{k-1} n_k = 571$, we have $n_1 n_2 n_3 = [(3 \cdot 10^3 - 4)/7] = 428$. Thus, $N = 428571$. If $n_{k-2} n_{k-1} n_k = 007$, then

$$G = \left\{ \frac{(3 \cdot 10^{567} - 4)}{571} \right\} \{007\} = 0052539n_8 \ldots n_{566}6007.$$

All 3-cycle generators are of the form:

$$G_0 = 0n_2 n_3 n_4 n_5 \ldots n_{k-3} \delta n_{k-1} n_k,$$
$$G_1 = 10n_3 n_4 n_5 \ldots n_{k-3} 0 n_{k-1} n_k,$$
$$G_2 = 110n_4 n_5 \ldots n_{k-3} 0 n_{k-1} n_k.$$

Note that $G_{i>0}$ exists only if $p > q$, whereas G_0 with $\delta = 1$ exists only if $p < q$. The search for 3-cycle generators can be further simplified since $(\delta n_{k-1} n_k, q \cdot 10^3 - p) = 1$ implies that $p \neq 5$ and $p.q \neq$ even.odd.

Distinct $p.q$ 3-cycles may have the same form. For example, consider

$$G_{(5.2)_{00}} = \left\{ (0n_{k-1} n_k) \cdot \left[\frac{(5 \cdot 10^{k-3} - 2)}{3 \cdot 5 \cdot 7 \cdot 19} \right] \right\} \{0n_{k-1} n_k\}.$$

Then $n_{k-1}n_k = 05$ yields

$$G_{(5.2)_{00}} = \left\{ \frac{5 \cdot 10^{15} - 2}{399} \right\} \{005\} = 012531328320802005.$$

Similarly, $n_{k-1}n_k = 35$ yields

$$H_{(5.2)_{00}} = \left\{ \frac{5 \cdot 10^{15} - 2}{57} \right\} \{035\} = 087719298245614035,$$

whereas $n_{k-1}n_k = 85$ gives

$$K_{(5.2)_{00}} = \left\{ \frac{5 \cdot 10^3 - 2}{399} \right\} \{005\} = 0238085.$$

It is easy to verify that $7 \cdot G_{(5.2)_{00}} = H_{(5.2)_{00}}$ and $19 \cdot G_{(5.2)_{00}} = \overline{K_{(5.2)_{00}}}^3$. In a similar fashion,

$$G_{(6.1)_{00}} = \left\{ \frac{6 \cdot 10^{207} - 1}{497} \right\} \{002\} = 012074n_8 \ldots n_{206}7002$$

and

$$H_{(6.1)_{00}} = \left\{ \frac{6 \cdot 10^{32} - 1}{71} \right\} \{014\} = 0845070n_8 \ldots n_{31}9014$$

satisfy $7 \cdot G_{(6.1)_{00}} = \overline{H_{(6.1)_{00}}}^6$. A partial sample of some other 3-cycle generators include:

$G_{(1.4)_{00}} = 0252n_6 \ldots n_{101}4101$

$G_{(2.5)_{00}} = 04081n_6 \ldots n_{38}5102$ $G_{(1.4)_{01}} = 02775n_6 \ldots n_{101}4111$

$G_{(1.3)_{00}} = 05152n_6 \ldots n_{995}5103$ $G_{(1.7)_{01}} = 017994n_7 \ldots n_{579}7111$

$G_{(4.7)_{00}} = 074615n_7 \ldots n_{22}2104$

$G_{(5.9)_{00}} = 058365n_7 \ldots n_{252}7105$ $G_{(5.4)_1} = 106382n_7 \ldots n_{42}8085$

$G_{(2.9)_{00}} = 023560n_7 \ldots n_{200}7106$ $G_{(7.4)_2} = 110443n_7 \ldots n_{238}6063$

$G_{(3.7)_{00}} = 0458768n_8 \ldots n_{1745}3107$

$G_{(4.7)_{00}} = 061749n_7 \ldots n_{22}9108$

$G_{(1.9)_{00}} = 012112n_7 \ldots n_{4492}9109$

With patience, one can easily produce all the 3-cycle generators and thus obtain all possible $p.q$ 3-cycles.

5. Conclusion. We have completely accounted for 1, 2, and 3-cycles. Although these results refer explicitly to numbers in base-ten notation, generalization to arbitrary bases also seems likely. We invite readers to probe into other properties of r-cycles, as well as to investigate $p.q$ r-cycles for $r \geqslant 4$.

REFERENCES

1. J. A. Hunter, Problem 908, Math. Mag., 48 (1975) 240–241.
2. S. Kahan, k-transposable integers, Math. Mag., 49 (1976) 27–28.
3. ———, k-reverse-transposable integers, J. Rec. Math., 9 (1976) 19–24.
4. ———, Rational-transposables, J. Rec. Math., 10 (1977-78) 163–168.
5. M. S. Klamkin, A number problem, Fibonacci Quart., 10 (1972) 324.
6. W. Page, $P.Q$ M-cycles: A generalized number problem, Fibonacci Quart., 12 (1974) 323–326.
7. ———, $P.Q$ 1-cycles: A number problem, Mathematics Student, Indian Math. Soc. (to appear 1982).
8. ———, Decimal digit shifting, The Two-Year College Math. Journal, 6 (1975) 25.
9. W. Shanks, On the number of figures in the period of the reciprocals of every prime number below 20,000, Proc. Royal Soc. London, 22 (1874) 200–210.
10. C. W. Trigg, A cryptarithm problem, Math. Mag., 45 (1972) 46.
11. J. Wlodarski, A number problem, Fibonacci Quart., 9 (1971) 195.

Coin Weighings Revisited

Aaron L. Buchman and Frank S. Hawthorne, *State University of New York, Albany, NY*

Perhaps you can solve the well-known folk problem:

> *A set of twelve seemingly identical coins contains one counterfeit coin. Using no more than three weighings on a two-pan balance beam, locate the counterfeit coin and tell if it is heavy or light.*

This problem aroused wide interest during the mid-1940's. Many ingenious proofs ([1], [2], [3]) were, and still continue to be, given of its generalization that

> *k ($k \geqslant 2$) weighings suffice to identify the counterfeit coin in any set of $n \in \{3, 4, \ldots, \frac{1}{2} \cdot (3^k - 3)\}$ coins.*

One of the more recent proofs of this generalization uses the following rather strange-sounding result due to B. Manvel [4].

Lemma. *Suppose a set of n coins contains a counterfeit coin, and each of the n coins is labelled "possibly heavy" (p.h.) or "possibly light" (p.l.). The least number of weighings needed to locate and identify the counterfeit coin is given by the unique number k satisfying $3^{k-1} < n \leqslant 3^k$.*

Proof. Suppose $n = 3^k$ (a similar proof holds when n is not a power of 3). Partition the set of n coins into three sets S_1, S_2, and S_3 of 3^{k-1} elements each. Put 3^{k-1} coins in the left pan, put 3^{k-1} coins in the right pan, and hold 3^{k-1} coins on the side. Make sure that each of the two pans contains an equal number of p.l. coins (and therefore an equal number of p.h. coins, as well). If the left pan is heavier, the counterfeit coin must be among the p.h. coins in the left pan or among the p.l. coins in the right pan—which collectively constitute a set of 3^{k-1} coins. If the scale balances, we are also left with 3^{k-1} coins to test. Therefore, in each case we reduced the size of the untested set of coins by a factor of 3. Continuing in this manner, we identify the counterfeit coin after the weighings.

275

The generalized folk problem may be rephrased as follows:

> *How many seemingly identical coins can be tested on a two-pan balance beam in order to determine whether the coins are equally weighted—and if not, to identify which is the unique heavy or light coin?*

That, of course, depends on the number of weighings allowed. Evidently three coins can be tested in two weighings, and up to twelve coins can be tested in three weighings. In general, we establish that

> k $(k \geqslant 2)$ *weighings suffice to test any set of*
> $n \in \{3, \ldots, \frac{1}{2} \cdot (3^k - 3)\}$ *coins.* $\qquad\qquad\qquad (*)$

Our proof is rather unusual since it actually produces an algorithm for easily locating and identifying the nature of a counterfeit coin (if it exists) when n is a multiple of three.

Begin by observing that there are three ways to employ any one of the n coins in each weighing. A coin's placement during the ith weighing $(1 \leqslant i \leqslant k)$ will be designated by $p_i \in \{0, 1, 2\}$, where

   0:   coin withheld from weighing in either pan,
   1:   coin placed in the right pan,
   2:   coin placed in the left pan.

And, at each weighing, the pan-balance assumes one of three positions $w_i \in \{0, 1, 2\}$ describing the state

   0:   both pans are level,
   1:   right pan is down,
   2:   left pan is down.

Thus, our sequence of k weighings determines both a k-digit base-three number $w_k w_{k-1} \ldots w_2 w_1$ describing the positions of the balance, and a k-digit base-three number $p_k p_{k-1} \ldots p_2 p_1$ monitoring where a coin was placed during this sequence of weighings. Our objective is to identify the counterfeit coin, if it exists, by matching its placement k-tuple $p_k p_{k-1} \ldots p_2 p_1$ with a uniquely determined descriptive balance k-tuple $w_k w_{k-1} \ldots w_2 w_1$. The following observations are basic:

(1) If two coins are kept together during each placement, we can at best determine that this pair contains the counterfeit coin—without knowing which coin it is. Therefore, no two coins may be assigned the same placement k-tuple. Since there are exactly 3^k distinct (placement) k-tuples with base 3 coordinates, it is necessary that $n \leqslant 3^k$.

(2) If all the coins are standard, then $w_i = 0$ for $1 \leqslant i \leqslant k$. The converse, however, tells us nothing if one of the coins is never placed in a pan.

Therefore, we require that every coin be weighed at least once. This means that no coin may be assigned the zero placement k-tuple $00 \ldots 0$, and so $n \leqslant 3^k - 1$. Note that $w_k w_{k-1} \cdots w_2 w_1 = 00 \ldots 0$ is now equivalent to the condition that all the coins are standard.

(3) If a coin is "heavy," its placement k-tuple $p_k p_{k-1} \cdots p_2 p_1$ must generate the same descriptive balance k-tuple $w_k w_{k-1} \cdots w_2 w_1$. If the coin is "light," its corresponding balance position k-tuple $w'_k w'_{k-1} \cdots w'_2 w'_1$ will be the reverse or "mirror" description of $w_k w_{k-1} \cdots w_2 w_1$; that is,

$$w_i = 1 \Leftrightarrow w'_i = 2 \quad \text{and} \quad w_i = 2 \Leftrightarrow w'_i = 1.$$

Thus, one "mirror" pair of k-tuples describing the balance positions must be reserved for each coin's placement k-tuple. Since there are $3^k - 1$ nonzero descriptive position k-tuples $w_k w_{k-1} \cdots w_2 w_1$, there are at most $\frac{1}{2}(3^k - 1)$ possible placement k-tuples assignable to our n coins. This necessitates that $n \leqslant \frac{1}{2}(3^k - 1)$.

(4) The upper bound $\frac{1}{2}(3^k - 1)$ on the number of coins that can be tested in k weighings is still too large. We claim that $n \leqslant \frac{1}{2}(3^k - 3)$. To see this, suppose that a set S of n coins is to be tested in k weighings. Partition S into three subsets: $S_1 = \{\text{coins in right pan}\}$, $S_2 = \{\text{coins in left pan}\}$, and $S_0 = \{\text{unweighed coins}\}$. If $w_1 = 0$, then [by (3) above] S_0 has cardinality $|S_0| \leqslant \frac{1}{2}(3^{k-1})$. If $w_1 \neq 0$, the coins in $S_1 \cup S_2$ must be testable in $k - 1$ weighings. Since all coins in $S_1 \cup S_2$ are p.h. or p.l., our Lemma yields $|S_1 \cup S_2| \leqslant 3^{k-1}$. However $|S_1| = |S_2|$, and so $2|S_1| \leqslant 3^{k-1}$. Since 3^{k-1} is odd, only $|S_1| + |S_2| = 2|S_1| \leqslant 3^{k-1} - 1$ is possible. Therefore,

$$n = |S| = |S_1| + |S_2| + |S_3| \leqslant (3^{k-1} - 1) + \frac{1}{2}(3^{k-1} - 1)$$

$$= \frac{3}{2}(3^{k-1} - 1) = \frac{1}{2}(3^k - 3).$$

Our proof of (∗) will be broken into two parts. First, we prove that:

$$n = \tfrac{1}{2}(3^k - 3) \text{ coins can be tested in } k(k \geqslant 2) \text{ weighings.} \tag{∗∗}$$

A few illustrations may be useful before turning to the proof of (∗∗).

Example. (a) Three coins can be tested in two weighings, using one coin in each pan. Verification: Let $\mathbf{1}_3$ consist of three coins, each of which is labelled by its placement 2-tuple $p_2 p_1$.

$$\mathbf{1}_3$$

$A = 01$	$B = 12$	$C = 20$

This can be shown diagrammatically as (the yet unweighed pans)

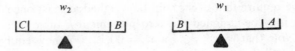

It is easy to verify that

$$w_2w_1 = 00 \Rightarrow A, B, C, \text{ are standard.}$$

$$w_2w_1 = 20 \Rightarrow C \text{ is heavy.} \quad (w_2w_1 = 10 \Rightarrow C \text{ is light.})$$

$$w_2w_1 = 01 \Rightarrow A \text{ is heavy.} \quad (w_2w_1 = 02 \Rightarrow A \text{ is light.})$$

$$w_2w_1 = 12 \Rightarrow B \text{ is heavy.} \quad (w_2w_1 = 21 \Rightarrow B \text{ is light.})$$

(b) Twelve coins can be tested in three weighings, using four coins in each pan. List these twelve coins as follows:

1_{3^2}		
$D = 001$	$E = 012$	$F = 020$
$G = 101$	$H = 112$	$I = 120$
$J = 201$	$K = 212$	$L = 220$

2_3		
$M = 011$	$N = 122$	$O = 200$

The members of 1_{3^2} are simply the numbers of 1_3, each of which was preceded by 0, 1, and 2. Clearly w_1 determines three elements in 1_{3^2} and one element in 2_3. And, w_2 further restricts the above choices to either the three elements of 1_{3^2} or the one element of 2_3. Finally, w_3 screens out one unique element in 1_{3^2}.

By way of further illustration, consider $w_3w_2w_1 = 120$. Diagrammatically,

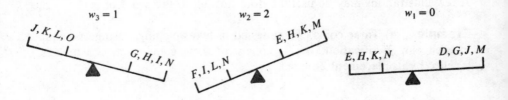

shows that

$$w_1 = 0 \Rightarrow \{D, E, G, H, J, K, M, N, \} \text{ are standard.}$$

$$w_2 = 2 \Rightarrow \{F, I, L\} \text{ contains a heavy coin.}$$

$$w_3 = 1 \Rightarrow I = 120 \text{ is the heavy coin.}$$

(c) Thirty-nine coins can be tested in four weighings, using thirteen coins in each pan. The thirty-nine coins can be listed as

$\mathbf{1}_{3^3}$			$\mathbf{2}_{3^2}$			$\mathbf{3}_3$		
0001	0012	0020	0011	0122	0200	0111	1222	2000
1001	1012	1020	1011	1122	1200			
2001	2012	2020	2011	2122	2200			
0101	0112	0120						
1101	1112	1120						
2101	2112	2120						
0201	0212	0220						
1201	1212	1220						
2201	2212	2220						

where the elements in $\mathbf{2}_{3^2}$ are the elements of $\mathbf{2}_3$, each preceded by 0, 1, and 2. At worst, four weighings are needed to single out a unique 4-tuple. This is clear since w_1 picks out one column of $\mathbf{1}_{3^3}$ (as well as one column in $\mathbf{2}_{3^2}$ and one element in $\mathbf{3}_3$); $w_2 \neq w_1$ gives no new information about this column in $\mathbf{1}_{3^3}$; w_3 selects three members of this column; and w_4 screens out the particular element $w_4 w_3 w_2 w_1 = p_4 p_3 p_2 p_1 \in \mathbf{1}_{3^3}$. Similar reasoning shows that $w_4' w_3' w_2' w_1'$ will appear if $p_4 p_3 p_2 p_1 \in \mathbf{1}_{3^3}$ is light.

It is a simple matter now to prove (∗∗) by induction. Assume that $n = \frac{1}{2}(3^k - 3)$ coins can be tested in k weighings for $k = 1, 2, \ldots, N$. We shall show that this formula holds for $k = N + 1$. By assumption, each of the $n = \frac{1}{2}(3^N - 3)$ coins may be assigned a unique N-tuple $p_N p_{N-1} \cdots p_2 p_1$ that results in this coin being selected as heavy (light) if $w_N w_{N-1} \cdots w_2 w_1$ (if $w_N' w_{N-1}' \cdots w_2' w_1'$) takes on the pattern $p_N p_{N-1} \cdots p_2 p_1$. These n coins are listed in sets

$$\mathbf{1}_{3^{N-1}} \quad \mathbf{2}_{3^{N-2}} \quad \cdots \quad (\mathbf{N-2})_{3^2} \quad (\mathbf{N-1})_3$$

The columns within each set contain an equal number of zeros, ones, and twos. There are no N-tuples consisting of all zeros, all ones, or all twos. We now assign $(N + 1)$-tuples in base three to our $\frac{1}{2}(3^{N+1} - 3)$ coins as follows: First, generate the set $\mathbf{1}_{3^N}$ by preceding each element in $\mathbf{1}_{3^{N-1}}$ by 0, by 1, and by 2. Since each column of $\mathbf{1}_{3^{N-1}}$ had 3^{N-2} zeros, ones, and twos, each column of $\mathbf{1}_{3^N}$ will contain exactly 3^{N-1} zeros, ones, and twos. Next, repeat this process with the other sets and obtain new sets

$$\mathbf{1}_{3^N} \quad \mathbf{2}_{3^{N-1}} \quad \cdots \quad (\mathbf{N-2})_{3^3} \quad (\mathbf{N-1})_{3^2}.$$

Finally, create the new set $\mathbf{N}_3$ consisting of three $(N + 1)$-tuples {011 ... 1; 122 ... 2; 200 ... 0}. The total number of $(N + 1)$-tuples so obtained is

$$3^N + 3^{N-1} + \cdots + 3^2 + 3 = 3\left[\frac{1 - 3^N}{1 - 3}\right] = \frac{1}{2}[3^{N+1} - 3].$$

It remains to show that $N + 1$ weighings will screen out and determine the nature of the counterfeit coin if it exists. Assume the worst; that is, assume that the counterfeit coin is in $\mathbf{1}_{3^N}$. Then w_1 selects one column of 3^{N-1} elements in $\mathbf{1}_{3^N}$. Clearly w_2 offers no new information. But w_3 reduces this column of 3^{N-1} elements to 3^{N-2} possibilities. And, w_4 further reduces these 3^{N-2} possibilities to 3^{N-3} possibilities. Proceeding in this manner, we see that w_{N+1} reduces the preceding $3^{N-(N-1)} = 3$ possibilities to the unique heavy coin having $p_{N+1}p_N \cdots p_2 p_1 = w_{N+1} w_N \cdots w_2 w_1$. The same type of argument assures that $p_{N+1}p_N \cdots p_2 p_1 = w'_{N+1} w'_N \cdots w'_2 w'_1$ when the counterfeit coin is light.

The proof of $(*)$ can now be completed by $(**)$, our Lemma, and induction.

In $k = 2$ weighings, $n = 3$ coins can be tested. In $k = 3$ weighings, $n \in \{4, \ldots, 12\}$ coins can be tested. We already know [by $(**)$] that 12 coins can be tested. Suppose $n \in \{4, \ldots, 11\}$ is written as $n = 2S + S_0$, where S_0 is chosen as the maximum possible value in $\{1, 2, 3\}$ for $S \in \{1, 2, 3, 4\}$. (Thus, we write $7 = 2(2) + 3$ rather than $7 = 2(3) + 1$.) Assume that S coins are placed in each pan and S_0 coins are unweighed during the first weighing. If $w_1 = 0$, only the S_0 coins require testing and this is easily accomplished in two weighings (by comparing with standard coins). If $w_1 \neq 0$, there are at most eight coins in the two pans, and these meet the conditions of our Lemma. Therefore, these eight coins can be tested in two weighings.

In $k = 4$ weighings, $n \in \{13, \ldots, 39\}$ coins can be tested. We know that 39 coins can be tested by $(**)$. Suppose $n \in \{13, \ldots, 38\}$ is written as $n = 2S + S_0$, where $S_0 \in \{1, 2, \ldots, 12\}$ is the maximum remainder for $S \in \{1, 2, \ldots, 13\}$. Assume that S coins are placed in each pan and keep S_0 coins unweighed during the first weighing. If $w_1 = 0$, the S_0 coins can be tested in $k = 3$ weighings. If $w_1 \neq 0$, there are at most twenty-six coins on the pans, and these can be tested in $k = 3$ weighings by our Lemma.

Suppose $n \in \{\frac{1}{2}(3^{k-1} - 1), \ldots, \frac{1}{2}(3^k - 3)\}$ coins can be tested in k weighings. We want to show that $n \in \{\frac{1}{2}(3^k - 1), \ldots, \frac{1}{2}(3^{k+1} - 3)\}$ coins can be tested in $(k + 1)$-weighings. We already know, by $(**)$, that $\frac{1}{2}(3^{k+1} - 3)$ coins can be tested in $(k + 1)$-weighings. Write

$$n \in \left\{ \tfrac{1}{2}(3^k - 1), \ldots, \tfrac{1}{2}(3^{k+1} - 5) \right\}$$

as $n = 2S + S_0$, where $S_0 \in \{1, 2, \ldots, \frac{1}{2}(3^k - 3)\}$ is the maximum remainder for $S \in \{1, 2, \ldots, \frac{1}{2}(3^k - 1)\}$. Assume that each pan has S coins, and S_0 coins are unweighed. Then $w_1 = 0$ insures that the S_0 unweighed coins can be tested in k weighings. If $w_1 \neq 0$, there are at most $3^k - 1$ coins in the two pans, and these can (by our Lemma) be tested in k weighings.

REFERENCES

The first three references are collectively written in an informal way by journal, author, and page number.

1. American Mathematical Monthly: 1945 (E. D. Shell, 42; M. Dernham, 397), 1946 (D. Eves, 156; N. J. Fine, 278), 1947 (E. D. Shell, 46; J. Rosenbaum, 48).
2. Mathematical Gazette: 1945 (R. L. Goodstein, 227), 1946 (F. S. Dyson, 231), 1947 (C. A. B. Smith, 31).
3. Scripta Mathematica: 1945 (H. D. Grossman, 360; L. Washington, Jr., 361), 1948 (C. W. Raine, 66; K. Itkin, 69).
4. B. Manvel, Counterfeit coin problems, Mathematics Magazine, 50 (March 1977) 90–92.

How to Tree Instant Insanity®

Jean J. Pedersen, *University of Santa Clara, Santa Clara, CA*

Introduction. The object of the popular "Instant Insanity®" puzzle (Parker Brothers) is to arrange the four cubes represented by the diagrams in Figure 1(a) in a $1 \times 1 \times 4$ stack so that a red, a green, a blue, and a white face appears on each of the four vertical sides of the stack (see Figure 1(b)). The ordering of the colors on the vertical faces is not of importance.

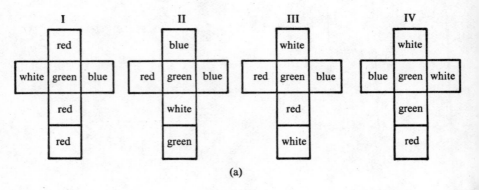

(a)

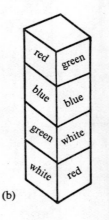

(b)

Figure 1.

282

The solution presented here uses tree diagrams and therefore is appropriate for standard courses in Mathematics for Liberal Arts, Finite Mathematics, etc. The method describes a way to deduce the solution to this and other similar puzzles in a straightforward logical manner that requires no advanced mathematical concepts. It proves, in this case, that the solution is unique (except for permutations of the cubes in the vertical stack, or rotations of the solved configuration in space). But, if there were no solution, or if there were multiple solutions, this method could also be used to identify those situations.

Initial Observation.   There are 24 faces on the four cubes: 7 red, 6 white, 6 green, and 5 blue faces. In the solved form, 16 faces (4 of each color) must be in a vertical position. The remaining eight horizontal faces are located either on the top of the stack, the bottom of the stack, or hidden between the cubes. Note that these faces are not faces that are used to satisfy the requirements for the solution. Accordingly, these eight faces will be called the *unimportant*, or *hidden* faces. Subtraction reveals that the 8 hidden faces must include 3 red, 2 white, 2 green and 1 blue face.

Plan of Attack.   First we determine which of the *opposing* faces on each cube are unimportant, and thus should be hidden. The hope is that this knowledge will substantially reduce the number of possibilities so that an exhaustive search of the remaining options will quickly yield the solution.

Let b, g, r, w represent the blue, green, red and white faces, respectively. We know from our initial observation that we need to identify pairs of opposing faces on the cubes such that the product of the assigned letter values on those faces is equal to $b^1 g^2 r^3 w^2$. From Figure 1 we see that the cubes have the following letters assigned to their opposing faces:

I	II	III	IV
rr	br	rw	gr
gr	bw	br	gw
bw	gg	gw	bw

Note that we record the opposing faces with the letters in alphabetical order until we have some reason to do otherwise.

To consider all combinations of the opposing faces, we draw a tree diagram as shown in Figure 2. Of course the diagram may be terminated when it becomes evident that a branch cannot lead to a solution. This is denoted in the figure by the symbol "|", followed by the reason for termination. For example: in the product along the top branch, the exponent of r^4 exceeds the permissible 3. Only five cases yield suitable combinations of hidden faces.

An Extension of the Original Plan.   Although we have considerably reduced the region of search, it is still not trivial (in terms of trial and error)

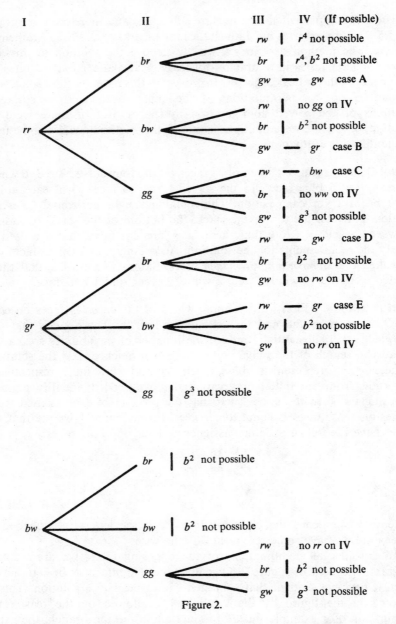

Figure 2.

to determine which, if any, of the cases A, B, C, D, E will lead to a solution of the original problem.

We must examine each of these cases. But now we need to look on each of the four cubes only at the two pairs of opposing faces which are not hidden. It might appear that we will have to check as many as 16 combinations for each of these cases. Fortunately, however, we can place the first cube with either available opposing pair of faces in a front-back orientation. Thus, we can choose a single pair of opposite faces from cube I and then examine all of the front-back combinations for the remaining three cubes. We can do this without loss of generality, since the solved puzzle can always be rotated 90 degrees about a vertical axis through the center of its horizontal faces. In these tree diagrams conflicts also occur, since we observe that in successful cases the product of the eight face values on the front and back of a satisfactory arrangement must be $(bgrw)^2 = b^2 g^2 r^2 w^2$.

Case A requires that we examine all combinations for the following situation:

Cube / face	I	II	III	IV	
hidden (horizontal)	rr	br	gw	gw	} This fixed row from branch A in Figure 2.
vertical	gr	bw	rw	gr	} The remaining opposite faces as
vertical	bw	gg	br	bw	} determined from Figure 1.

Figure 3 readily verifies that Case A cannot lead to a solution of the puzzle. The reader may wish to verify, in a similar manner, that cases B, D and E cannot lead to a solution.

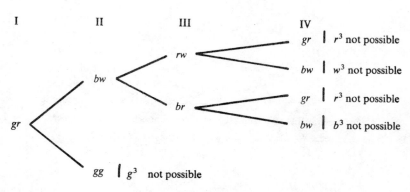

Figure 3.

It is case C that leads to a unique solution. This is seen in Figure 4 which represents a tree diagram for the combinations of the cubes appropriately categorized thus:

Cube face	I	II	III	IV	
hidden (horizontal)	rr	gg	rw	bw	This fixed row from branch C in Figure 2.
vertical	gr	br	br	gr	The remaining opposite faces as determined from Figure 1.
vertical	bw	bw	gw	gw	

To find the solution from the information in Figure 4 notice that the upper "possible" branch, as represented now, designates the *front* faces as g, b, b, g. Clearly this is not satisfactory. But don't give up. Observe that the tree diagram would be just as valid if you exchange any of the individual pairs of letters at its branch points. (This is equivalent to rotating that particular cube 180 degrees about the horizontal axis of symmetry passing through the center of the left-right vertical faces.) Thus the branch now labeled gr—bw—br—gw might be rearranged, by the exchange of letters in the third and fourth branch points, to obtain gr—bw—rb—wg. But then the colors b, g, r, w would appear exactly once as a first element and exactly once as a second element in that branch.

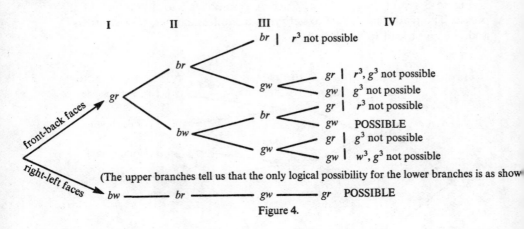

Figure 4.

Likewise an exchange of letter values in the first and fourth branch points of the "complementary" lower "possible" branch, which now appears as bw—br—gw—gr, would give wb—br—gw—rg.

These two branches, along with the information already known about the hidden faces yields the following description of the unique solution to this puzzle:

Cube faces	I	II	III	IV
hidden (horizontal)	rr	gg	rw	bw
front-back (vertical)	gr	bw	rb	wg
right-left (vertical)	wb	br	gw	rg

} This row from case C in Figure 2.

} These values obtained by a suitable adjustment of the two "possible" branches in Figure 4.

You may have observed that adjusting the lower "possible" branch to *bw* —*rb*—*wg*—*gr* would have led to an apparently different solution. However, this solution can be obtained from the first one by rotating the entire solved puzzle first through 180 degrees about its vertical axis of symmetry and then 180 degrees through its right-left horizontal axis of symmetry. Other apparent variations can, likewise, be shown to be equivalent to the solution cited.

Appreciation. If you would like an instructive problem, label four cubes with colors of your own choice and use the method described here to determine how many solutions are possible. You will then have a greater appreciation for this particular puzzle.

A Note for Computer Programmers. Since this solution does not ever require that a judicious, or intuitive, choice be made on the part of the solver, it represents an algorithm for which a computer could be programmed to solve the puzzle. If you were to write such a program, you could use it to find solutions for the puzzles you devise similar to the one given here.

Post Script About a Similar Puzzle. In Leslie E. Shader's article, "Cleopatra's Pyramid" [*Mathematics Magazine* (January, 1978), 57–60], it is correctly asserted that the number of different arrangements of four unit cubes (each face of which is one of four colors) into a $1 \times 1 \times 4$ unit stack is $2(12)^4$. Likewise, the correct number of arrangements of four unit regular tetrahedra and a one unit regular octahedron (each triangular face of which is one of four colors) into a two unit regular tetrahedron was computed to be $4(12)^5$ (larger than the first arrangement by a factor of 24). The completed arrangements are illustrated in Figure 5.

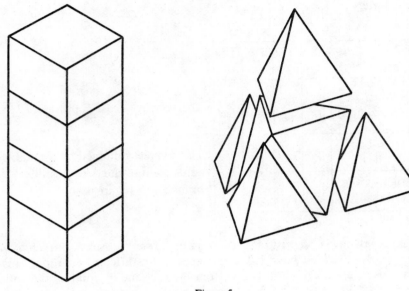

Figure 5.

In Professor Shader's article, the tetrahedral analogue of Instant Insanity® is developed, and it is shown that the "best" coloring scheme (that is, the one giving the *least* number of possible solutions) is the arrangement of colors given on the nets of Figure 6. To successfully complete the puzzle, it is required that all four colors be visible on each of the four faces of the large tetrahedron that is assembled from these five polyhedra.

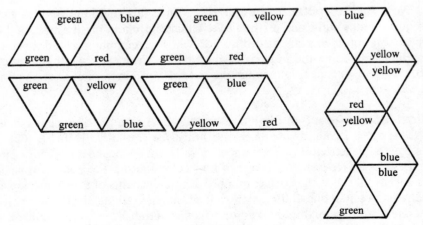

Figure 6.

This puzzle, in contrast with Instant Insanity® which as we've seen has a *unique* solution, has 19 different solutions. However, an alternate puzzle, using the same five polyhedra and requiring that each face of the large tetrahedron be *monochromatic* does have a unique solution.

It is asserted, in Shader's article, that the tetrahedral puzzle requiring the four colors on each face would appear to be 24/19 times as difficult as Instant Insanity®. By the same reasoning the monochromatic version of this puzzle would appear to be 24/1 times as difficult as Instant Insanity®.

A Surprisingly Simple Solution. Consider the following solution to the monochromatic version of Cleopatra's Pyramid. For clarity, *triangle* will refer to any 3-sided polygon on one of the original five puzzle pieces, and "face" (including the quote marks) will refer to a 3-sided polygon (made up of four triangles) on the completed puzzle.

(1) Observe that the five puzzle pieces include 24 triangles of which

> 8 are green,
> 6 are blue,
> 6 are yellow,
> 4 are red.

Eight of these triangles must be hidden on the completed puzzle.

(2) Since there is only one green triangle on the octahedron, we know that there cannot be two green "faces" on the completed puzzle (otherwise, the 8 green triangles might have admitted that possibility). Thus, the four "faces" must include all four colors.

(3) Because each "face" must be monochromatic, we see that no red triangle can be hidden. Since there is only one red triangle on the octahedron, its set of four hidden triangles (no two of which share an edge) is then specified; it includes two blue and two yellow triangles.

(4) We now know exactly which four triangles are hidden on the octahedron, and we know that the following triangles remain:

> 8 green,
> 6 − 2 = 4 blue,
> 6 − 2 = 4 yellow,
> 4 red.

Since all of the unspecified blue, yellow, and red triangles must show, and since the green triangle on the octahedron cannot be hidden, we deduce that when each of the four tetrahedra are placed on the octahedron, it must be done so as to hide a green triangle on that tetrahedron.

(5) The above information, coupled with the knowledge that each "face" is monochromatic, enables one to solve the puzzle quickly by trial and error. Furthermore, since no tetrahedron can be placed satisfactorily on more than one triangular face of the octahedron, it is verified that this solution is unique.

It is interesting to recall that (according to a perfectly logical mathematical argument) Cleopatra's Pyramid, in the monochromatic version, was seen to be 24 times as difficult as Instant Insanity®. Nevertheless, this seemingly more difficult puzzle yields to a remarkably direct and simple solution. No such easy and concise solution is yet known for Instant Insanity®. A monochromatic arrangment of the four vertical faces of Instant Insanity® is easily seen to be impossible. (To see why, look carefully at the faces surrounding the three axes of symmetry joining opposite faces on cube number II.)

The symmetry of the octahedron and its relationship to the four attached tetrahedra would seem, at first glance, to add enormous difficulty to the puzzle. However, an examination of the solution reveals that exactly the opposite occurs. Thus, assigning a "difficulty index" to puzzles of this type is not at all a simple matter.

I wish to thank Franz O. Armbruster for talking to the students in my Mathematics for Liberal Arts course about this puzzle—most notably, for explaining his role in popularizing it and for relating the fascinating solution using a television scanner, a mechanical arm, and a computer that was done at the Artificial Intelligence Center at Stanford. His suggestion that it might be useful to look at the "hidden" faces was the underlying concept that led to the solutions described in this paper. I also thank Professor Dave Logothetti for suggesting the title of this article.

REFERENCES

1. Robert G. Busacker and Thomas L. Saaty, Finite Graphs and Networks, McGraw-Hill, New York, 1965.
2. T. A. Brown, A note on "Instant Insanity," Math. Mag., 41 (1968) 167–169.
3. J. V. Devertes, The Many Facets of Graph Theory, Lecture Notes in Mathematics, 110, Springer-Verlag, New York, 1968, p. 283.
4. Dewey D. Duncan, Instant Insanity: that ubiquitous baffler, Math. Teacher, 65 (February 1972) 131-135. (This solution utilizes the tree concept.)
5. A. P. Grecos and R. W. Gibberd, A diagrammatic solution to "Instant Insanity" problem, Math. Mag., 44 (1971) 119–124.
6. T. H. O'Beirne, Puzzles and Paradoxes, Oxford Univ. Press, New York, 1965, 112–129. This book gives a history of puzzles of this type, and suggests methods of solution that generally require some judicious choices at some point.
7. William L. Schaaf, A Bibliography of Recreational Mathematics NCTM, vols. 2 [no. 135, 1970, pp. 144–145], 3 [no. 105, 1973, pp. 83–84] and 4 [no. 246, 1978, pp. 99–100]. (Vols. 2 & 3 list references up to 1973, and vol. 4 lists references since 1972.)
8. Leslie E. Shader, Cleopatra's pyramid, Math. Mag., 51, no. 1 (1978) 57–60.
9. B. L. Schwartz, An improved solution to Instant Insanity, Math. Mag., 43 (1970) 20–23. (This solution utilizes the exponent idea.)

The Game Beware

F. Pedersen and K. Pedersen, *Southern Illinois University, Carbondale, IL*

This article presents a game, called Beware, that was motivated by the study of generalized fixed-point theory. While the game is self-contained, it can be used to illustrate mathematical concepts such as topology, continuity, paths, and trees.

The rules for playing Beware are given in Section I. No knowledge of mathematics is needed for this section. Section II introduces some mathematical concepts and illustrates their relationship with the game Beware. Section III contains the proofs of the theorems which motivated the construction of Beware, and Section IV offers suggestions for using Beware to explore other related mathematical concepts.

I. How to Play Beware

The playing board consists of a surface on which is drawn one of the following designs:

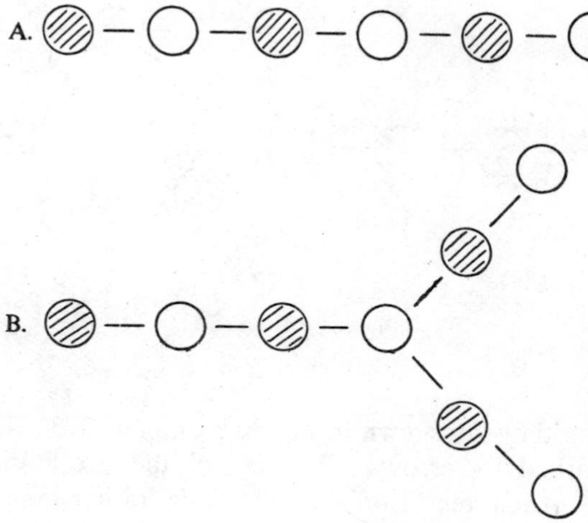

Figure 1.

The first player, designated the "trapper," is allowed to choose the number of markers to be played. The trapper then places his set of red markers $\{1, 2, \ldots, n\}$ on the board according to the following rules:

R_1: Odd-numbered markers must be placed on shaded circles.

R_2: Even-numbered markers must be placed on nonshaded circles.

R_3: Adjacent-numbered markers must be placed on adjacent circles.

R_4: The markers must be placed in succession; that is, marker number 1 is played; then marker number 2 is played; then marker number 3 until all n markers have been placed on the board.

After the trapper has placed his n red markers on the board, the other player, the "fox," places his set of black markers $\{1, 2, \ldots, n\}$ on the same board according to rules R_1–R_4 above.

The object of the game is for the fox to place all n of his black markers on the board without putting a numbered marker on a circle on which the trapper has already placed the same red-numbered marker. (Neither player may move a marker after it has been played.) If the fox's marker matches the trapper's on some circle, the game is over and the trapper wins. Otherwise, the fox wins.

Example of a played game. Assume that board design B has been chosen and that the trapper decided to play with 15 markers. The trapper's play is shown in Figure 2. (For ease in reading, numbers below circles indicate plays of the trapper and numbers above the circles indicate plays of the fox.)

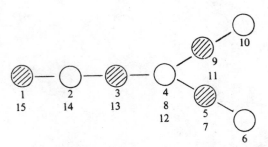

Figure 2.

The fox responds with the play shown in Figure 3 on page 293. The fox has lost! (Note play marked by arrow.) The play of the fox at marker 12 necessitated the play of marker 13 on the first circle, rather than the third circle, because the trapper had already placed number 13 on the third

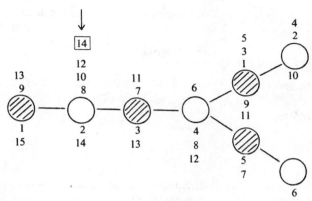

Figure 3.

circle. Thus, the fox had to play number 14 on the second circle. But the trapper had already played his number 14 on the second circle.

Variations of the rules. A *proper move* is any move in accordance with rules R_1–R_4. The rules may be changed to permit either player to play an odd-numbered marker on a nonshaded circle or an even-numbered marker on a shaded circle if

(a) he has first made a proper move on the circle in question, and

(b) he does not play on another circle except by a proper mover.

An example of this variation on a play by the trapper is seen in Figure 4. Board design A is chosen.

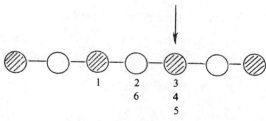

Figure 4.

The arrow indicates the use of the variation. Marker number 4 can be placed on a shaded circle since the placement of marker number 3 was a proper move. Marker number 5 is placed on this same circle since the trapper can leave this circle only by a proper move. The placement of marker number 6, an even-numbered marker, on the adjacent nonshaded circle is a proper move.

II. Topological Concepts

This section serves as the prelude to a mathematical interpretation of the game Beware. Here we introduce and illustrate some elementary notions in point set topology.

1. Topological Spaces

A *topology* on a set X is a family $\mathfrak{T}$ of subsets of X (called *open*) that contains $\emptyset$, X and is closed under finite intersections and arbitrary unions. The complement $X\text{-}E$ of (an open set) $E \in \mathfrak{T}$ is called a *closed* set. Thus $\emptyset$, X are both open and closed. The pair $(X, \mathfrak{T})$ is termed a *topological space*—abbreviated "*space*" when no confusion is likely.

If Y is a subset of $(X, \mathfrak{T})$ then $\mathfrak{T} \cap Y = \{ E \cap Y : E \in \mathfrak{T} \}$ is a topology on Y. Accordingly, $(Y, \mathfrak{T} \cap Y)$ is termed a *subspace of* $(X, \mathfrak{T})$.

One standard technique for specifying a topology on a set is as follows.

Let $\mathfrak{U} = \{ u_\alpha : \alpha \in A \}$ be a family of subsets of X that satisfies:

For each $u_\alpha, u_\beta \in \mathfrak{U}$ and $x \in u_\alpha \cap u_\beta$, there is a $u_\gamma \in \mathfrak{U}$ such that
$$x \in u_\gamma \subset u_\alpha \cap u_\beta. \qquad (*)$$

Then $\mathfrak{T}(\mathfrak{U}) = \{ \emptyset, X,$ all unions of members of $\mathfrak{U} \}$ is the smallest topology on X that contains $\mathfrak{U}$. Thus, $\mathfrak{U}$ completely determines $\mathfrak{T}(\mathfrak{U})$.

A space X is to be T_0 if for each pair of distinct points in X, there is an open set containing one of the points and not the other.

For the purpose of this paper, all spaces are assumed to be finite (i.e., X is finite) and T_0.

2. Continuous Functions and Homeomorphisms

A mapping $f : X \to Y$ between two spaces X and Y is said to be:

(i) *open*[†] (*closed*) if the image of each X-open (closed) set is an open (closed) set in Y,

(ii) *continuous* if the inverse of each open set in Y is an open set in X.

Remark. Since $f^{-1}(Y - E) = X - f^{-1}(E)$ for every open set $E = Y$, "open" can be replaced by "closed" in (ii).

Remark. If Y carries $\mathfrak{T}(\mathfrak{U})$ for $\mathfrak{U}$ as in $(*)$, then f is continuous if and only if $f^{-1}(u)$ is open in X for each $u \in \mathfrak{U}$. This is clear since f^{-1} preserves

---

[†] If f is bijective (i.e., one-to-one and onto), then being open is equivalent to being closed.

unions [i.e., $f^{-1}(\cup_A u_\alpha) = \cup_A f^{-1}(u_\alpha)$], and every $E \in \mathcal{T}(\mathcal{U})$ is the union of members of $\mathcal{U}$.

A continuous, open (closed) bijection is called a *homeomorphism*.

3. Connected Spaces

A space X is said to be *connected* if it is not the union of two nonempty disjoint open subsets. Evidently X is connected if and only if $\emptyset$ and X are the only subsets of X that are both open and closed.

Since inverse mappings preserve intersections as well as unions, the continuous image of a connected space is connected.

4. Examples

(a) On any set X, we can define topologies $\mathcal{D} = \{$all subsets of $X\}$ and $\mathcal{I} = \{\emptyset, X\}$. The space $(X, \mathcal{I})$ is connected and non-T_0, whereas $(X, \mathcal{D})$ is T_0 and nonconnected. Every mapping $f: (X, \mathcal{D}) \to Y$ from $(X, \mathcal{D})$ to any space Y is continuous. Note that the continuous bijection $1: (X, \mathcal{D}) \xrightarrow[x \to x]{} (X, \mathcal{I})$ is not open, and therefore a nonhomeomorphism.

(b) Let $\mathcal{T}_n$ and $\mathcal{T}'_n$ be topologies for $X_n = \{1, 2, \ldots, n\}$ determined as follows: For each integer $k \in \{1, 2, \ldots, n\}$, define

$$u_k = \begin{cases} \{k\}, & k \text{ odd}, \\ \{k-1, k, k+1\} \cap X_n, & k \text{ even}, \end{cases}$$

and

$$u'_k = \begin{cases} \{k-1, k, k+1\} \cap X_n, & k \text{ odd}, \\ \{k\}, & k \text{ even}. \end{cases}$$

Then $\mathcal{V}_n = \{u_k : 1 \leqslant k \leqslant n\}$ and $\mathcal{V}'_n = \{u'_k : 1 \leqslant k \leqslant n\}$ satisfy (∗), and we define $\mathcal{T}_n$ as $\mathcal{T}(\mathcal{V}_n)$ and $\mathcal{T}'$ as $\mathcal{T}(\mathcal{V}'_n)$. For convenience, I_n denotes the space $(X_n, \mathcal{T}_n)$ and I'_n denotes $(I_n, \mathcal{T}'_n)$.

Both spaces I_n and I'_n are connected, T_0 spaces. For instance, suppose $I_n = A \cup B$ for nonempty disjoint open sets $A, B \subset I_n$. Then $k \in A$ and $k + 1 \in B$ (or vice versa) for some $k \in I_n$. Since u_k is the smallest open set containing k, we have $u_k \subset A$. Similarly, $u_{k+1} \subset B$. This leads to the contradiction that $u_k \cap u_{k+1} = \emptyset$.

It is easily verified that $h: I_4 \xrightarrow[k \to 5-k]{} I'_4$ is a homeomorphism. On the other hand, I_5 and I'_5 are nonhomeomorphic since no bijection can map the open sets of $\mathcal{U}_5 = \{\emptyset, X_5, \{1\}, \{1, 2, 3\}, \{3\}, \{3, 4, 5\}, \{5\}\}$ on a one-to-one basis with the open sets of $\mathcal{U}'_5 = \{\emptyset, X_5, \{1, 2\}, \{2\}, \{2, 3, 4\}, \{4\}, \{4, 5\}\}$. Accordingly, there is no one-to-one correspondence between the open sets of $\mathcal{T}_5$ and those of $\mathcal{T}'_5$.

In general, I_n and I'_n are homeomorphic for even n, and non-homeomorphic for odd $n > 1$.

(c) Each set A and B below may be topologized by defining u_0 to be the circle $\bigcirc$ (i.e., "point" in $X = A$ or B) itself if the circle is unshaded, and by defining $u_\bullet$ to be the circle ● with its adjacent unshaded "points" in $X(= A$ or $B)$ if the circle is shaded.

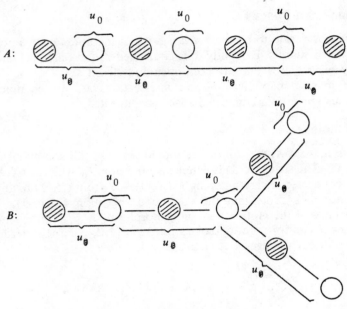

Figure 5.

The collection $\mathcal{U} = \{u_\bigcirc, u_\bullet : \bigcirc, ● \in X\}$ satisfies ($*$). Thus A and B are assumed to carry $\mathcal{T}(\mathcal{U})$.

Both A and B are connected, T_0 spaces. This is clear for A since

$$h : \quad I_7' \longrightarrow A$$

$$k \to k\text{th circle from the left}$$

is a homeomorphism. The reader can easily verify that B is connected, T_0 by considering the definition of $\mathcal{T}(\mathcal{U})$ on B. It may be instructive to show that B is nonhomeomorphic to I_8 (or I_8'). First, note that the only connected sets in I_n are the intervals $\{m \in I_n : j \leqslant m \leqslant k\}$. Let x denote the point (i.e., circle) of B where the three branches meet. There are three 3-point sets (call them E, F, G) which contain x and are both open and connected. If there were a homeomorphism $h : B \to I_8$, then $h(x)$ would be a point of I_8 that is contained in three 3-point sets $\{h(E), h(F), h(G)\}$ each of which is both open and connected. But there is no such point in I_8.

5. Paths and Trees

A nonempty subset E of a space X is called a *path* if E is homeomorphic to I_n or I_n' for some $n \geqslant 2$. A *path from x to y* $(x, y \in X)$ is the image $\mathbf{P}(x, y)$

of a homeomorphism $h : J_n (= I_n$ or $I_n') \to \mathbf{P}(x, y)$ satisfying $x = h(1)$ and $y = h(n)$.

Note. Since homeomorphisms are bijective, paths must consist of distinct (non-repeating) point images of members in J_n.

A *tree* is a T_0-space T satisfying:

(i) Each pair of points $x, y \in T$ is joined by a unique path $\mathbf{P}(x, y) \subseteq T$.

(ii) If $\mathbf{P}(x, y) \cap \mathbf{P}(y, z) = \{y\}$ $(x, y, z, \in T)$, then $\mathbf{P}(x, y) \cup \mathbf{P}(y, z) \subseteq T$ is a path from x to z.

The spaces I_n and I_n' are trees for every natural number n. Paths in A (and B), are sequences of adjacent points (i.e., circles joined by a dash in Figure 1) that do not overlap. Therefore, A and B are trees. Other examples of trees are shown below.

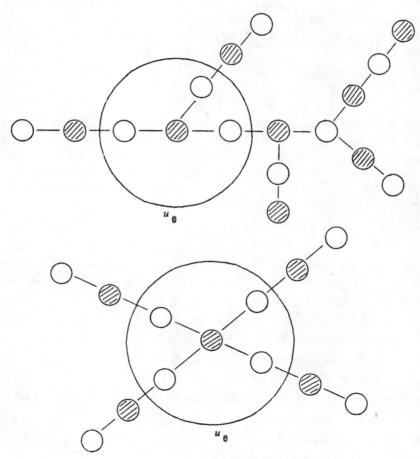

Figure 6.

6. Universal Function and Universal Images

A function $f: X \to X$ is said to have a *fixed point* if there is an $x_0 \in X$ such that $f(x_0) = x_0$. (Equivalently, there is an $x_0 \in X$ such that $f(x_0) = 1(x_0)$, where $1: X \underset{x \leadsto x}{\to} X$ is the identity map.) A space X has the *fixed-point property* if every continuous function $f: X \to X$ has a fixed point.

Each of these fixed-point concepts may be generalized: A continuous function $f: X \to Y$ is a *universal function* if for any continuous function $g: X \to Y$, there is an $x \in X$ such that $f(x) = g(x)$. A connected space Y is a *universal image* if $f: X \to Y$ is a universal function for every connected space X and continuous, onto function f.

Example 6a. We show that I_n is a universal image for each natural number n. A similar proof can be given for I'_n.

Let X be a connected space, and let $f: X \to I_n$ be a continuous mapping of X onto I_n. We must show that if $g: X \to I_n$ is any continuous mapping, then $g(x) = f(x)$ for some $x \in X$. Assume, to the contrary, that $g(x) \neq f(x)$ for every $x \in X$. Let

$$E = \{x \in X : f(x) < g(x)\} = \{x \in X : f(x) \leqslant g(x)\}$$

and

$$F = \{x \in X : f(x) > g(x)\} = \{x \in X : f(x) \geqslant g(x)\}.$$

Then $X = E \cup F$ and $E \cap F = \emptyset$. Since f is onto, there exist points $x, x' \in X$ for which $f(x) = 1$ and $f(x') = n$. Clearly, $x \in E$ and $x' \in F$. Hence neither E nor F is empty. We will show that E and F are both open, thereby contradicting the assumption that X is connected. For any element $x_0 \in E$, we show that there exists an open set u such that $x_0 \in u \subset E$. Let $r = f(x_0) < g(x_0) = s$.

Case 1 (r and s both odd). The set $u = f^{-1}(r) \cap g^{-1}(s)$ is open, contains x_0, and is contained in E.

Case 2 (r and s both even). In this case, $s - r \geqslant 2$ and $s - 1 \geqslant r + 1$. The set $u = f^{-1}(\{r - 1, r, r + 1\} \cap I_n) \cap g^{-1}(\{s - 1, s, s + 1\} \cap I_n$ is open, contains x_0, and is contained in E.

Case 3 (r even and s odd). The set $u = f^{-1}(\{r - 1, r, r + 1\} \cap I_n) \cap g^{-1}(s)$ is an open set in E that contains x_0.

Case 4 (r odd and s even). Similar to Case 3.

In similar fashion, F can be shown to be open.

The following characterization of finite T_0 universal images can be found in [2].

Theorem 1. *Every n-point universal image is homeomorphic to I_n or I'_n.*

The remarks in Example 4(b) can be used to show that, up to a homeomorphism, there exists only one n-point universal image when $n = 1$

or n is even, and there exist exactly two nonhomeomorphic n-point universal images when $n > 1$ is odd.

7. An Overview of the Game Beware

Let $t: I'_n \to Y$ and $f: I'_n \to Y$ represent the respective plays of the trapper and the fox from their set of markers I'_n to the playing board $Y (= A$ or $B)$. The topologies on I'_n and Y make these spaces trees. It will be shown, in Section III, that t and f are continuous. The objective of Beware may be viewed as asking if there is some marker $k \in \{1, 2, \ldots, n\}$ such that $t(k) = f(k)$. Answers and playing strategies will be given in the next section.

III. Mathematical Theory Behind Beware

This section provides a detailed exposition of the mathematics that motivated the game Beware. We begin by defining conditions under which a two-point set in a tree is a path. Notationwise (as in Examples 4(b), (c) of Section II), u_x denotes the smallest open set containing x.

Theorem 2. *Let* $x, y \in T$, *where* T *is a tree. Then* $\{x, y\}$ *is a path in* T *from* x *to* y *if and only if* $x \in u_y$ *or* $y \in u_x$.

Proof. If $\{x, y\}$ is a two-point path, it is homeomorphic to I_2 or I'_2. Therefore $\{x, y\}$ is connected, and either $x \in u_y$ or $y \in u_x$. Suppose conversely that $x \in u_y$. Since T is a T_0-space and u_y is the smallest open set about y, we have $y \notin u_x$. Furthermore, $x \in u_y$ implies that $u_x \subset u_y$. Therefore $\{x, y\}$ with the subspace topology $\{\emptyset, \{x\}, \{x, y\}\}$ is the homeomorph of I_2 or I'_2.

Remark. There exists a path $\{x, y\}$ if and only if there is a path $\{y, x\}$. To be sure, suppose $h: I_2 \to \{x, y\}$ with $h(1) = x$ and $h(2) = y$ is a homeomorphism, and let $h': I'_2 \to I_2$ satisfy $h'(1) = 2$ and $h'(2) = 1$. Then $h \circ h': I'_2 \to \{x, y\}$ is a homeomorphism with $(h \circ h')(1) = y$ and $(h \circ h')(2) = x$. ∎

When is a subset of a tree a tree under the subspace topology?

Theorem 3. *A subset* S *of a tree* T *is a tree if and only if* S *is connected.*

Proof. We first show that trees are connected. Suppose, to the contrary, that $T = A \cup B$, where A and B are disjoint nonempty open sets. Let $a \in A$ and $b \in B$, and let $P(a, b)$ be the path from a to b. Then $P(a, b) = \{A \cap P(a, b)\} \cup \{B \cap P(a, b)\}$ is the disjoint union of open sets in $P(a, b)$. This however is impossible since paths are connected. Thus, T is connected. Accordingly, S is connected if S is a tree.

Assume that S is connected. We shall show that S satisfies 5(i), (ii) of Section II. Fix any $x \in S$ and define $E_x = \{y \in S : \text{there is a path } P(x, y) \subseteq S\}$. Our objective is to establish that $E_x = S$. For any $y \in E_x$, let

$v_y = u_y \cap S$ be the open set (in the subspace topology) about $y \in S$. If $v_y = \{y\}$, then $v_y \subset E_x$. If $v_y \neq \{y\}$, there exists an element $z \in (v_y - \{y\}) \cap S$. Since $z \in u_y$, there is (by Theorem 2) a path $\{z, y\}$ in T from z to y, and a path $\{y, z\}$ in T from y to z. Since $y, z \in S$ and $\{y, z\}$ is a two-point path, $\{y, z\} \subset S$. If $z \in P(x, y)$, there is a path in S from x to z, and so $z \in E_x$. If $z \notin P(x, y)$, then $P(x, y) \cap \{y, z\} = \{y\}$, and so $P(x, y) \cup \{y, z\}$ is a path in S from x to z. Thus, $z \in E_x$. Since this shows that $z \in E_x$ for each $z \in (v_y - \{y\}) \cap S$, we see that $v_y \subset E_x$. Therefore each $y \in E_x$ is contained in an S-open set $v_y \subset E_x$, and E_x is open in S since it is the union of open sets in S.

To show that E_x is closed in S (and therefore conclude that $E_x = S$), we prove that $S - E_x$ is an open set in S. Consider any $y \in S - E_x$. We claim that there is some open set $u(y) \subset S$ such that $u(y) \cap E_x = \emptyset$. Therefore $y \in u(y) \subset S - E_x$ and the openness of $S - E_x$ in S will follow. To prove our claim, assume that $u(y) \cap E_x \neq \emptyset$ for every open set $u(y) \subset S$ which contains y. Then $v_y = u_y \cap S$ satisfies $v_y \cap E_x \neq \emptyset$. Let $z \in v_y \cap E_x$. Then (by Theorem 2) $z \in u_y$ implies that $\{z, y\}$ is a path in T from z to y, and $\{y, z\}$ is a path in T from y to z. Since $y, z \in S$, we have $\{y, z\} \subset S$. At the same time, $z \in E_x$ insures that there is a path $P(z, x) \subset S$. Therefore $\{y, z\} \cup P(z, x)$ is a path in S from y to x. This means that $y \in E_x$, and the definition of y is contradicted.

Since S is connected, and E_x is both open and closed, we have $S = E_x$. This being true for each $x \in S$, it is clear that S satisfies 5(i), (ii) or Section II. ∎

The astute observer may have already noticed that each point in the game board $Y(= A \text{ or } B)$ is either an open or closed set. Our next result shows why this must be true.

Theorem 4. *If T is a tree with at least two points and $x \in T$, then $\{x\}$ is either open or closed, but not both.*

Proof. If $u_x = \{x\}$, then $\{x\}$ is open. Assume that $u_x \neq \{x\}$. To show that $\{x\}$ is closed, we prove that $T - \{x\}$ is open. Suppose $y \neq x$. We prove that $x \notin u_y$, and so $y \in u_y \subset T - \{x\}$. Assume, to the contrary, that $x \in u_y$. Then $y \notin u_x$ since T is a T_0-space. Since $u_x \neq \{x\}$, there is some $z \neq x$ in u_x and (again since T is T_0) $x \notin u_z$. Theorem 2 shows that $\{x, y\}$ and $\{z, x\}$ are two-point paths in T. Since $\{z, x\} \cap \{x, y\} = \{x\}$, it follows that $\{z, x, y\} = \{z, x\} \cup \{x, y\}$ is a path in T from z to y. Accordingly, there is a homeomorphism

$$h : I_3 \to \{z, x, y\} \quad \text{with} \quad h(1) = z, h(2) = x, h(3) = y,$$

or a homeomorphism

$$h' : I_3' \to \{z, x, y\} \quad \text{with} \quad h'(1) = z, h'(2) = x, h'(3) = y.$$

Since $u_3 \cap I_3 = \{3\}$, the first case yields $u_y \cap \{z, x, y\} = \{y\}$, and this contradicts our assumption that $x \in u_y$. Since $u_2' \cap I_3' = \{2\}$, the second case yields $u_x \cap \{z, x, y\} = \{x\}$, and this contradicts $z \in u_x$. These contradictions establish that $x \notin u_y$ as asserted.

Finally, $\{x\}$ cannot be both open and closed because T is connected (Theorem 3). ∎

The following definition will be useful:

Distinct points x, y in a tree T are said to be *adjacent* if they form a two-point path $\{x, y\} \subset T$.

The property of being adjacent is not quite invariant under continuous mappings.

Theorem 5. *Let f be a continuous mapping from tree T_1 to tree T_2, and suppose $x, y \in T_1$ are adjacent. Then $f(x), f(y) \in T_2$ are either adjacent or identical.*

Proof. Since $\{x, y\} \subset T_1$ is connected, the set $\{f(x), f(y)\} \subset T_2$ is connected. If $f(x) \neq f(y)$, then $\{f(x), f(y)\}$ is a connected two-point set. Therefore either $f(x) \in u_{f(y)}$ or $f(y) \in u_{f(x)}$. In particular, $\{f(x), f(y)\}$ is a path since it is homeomorphic to I_2. ∎

The preceding result can be put into sharper focus.

Theorem 6. *Let f be a continuous mapping from tree T_1 to tree T_2, and suppose $x, y \in T_1$ are adjacent. If $f(x) \neq f(y)$, then $\{x\}$ and $\{f(x)\}$ are either both open or both closed.*

Proof. Suppose $\{x\} \in T_1$ is closed and $\{f(x)\} \in T_2$ is open. Then (see 2(ii) of section II) $f^{-1}\{f(x)\} \subset T_1$ is open and contains x. By definition, $u_x \subset f^{-1}\{f(x)\}$. Since y is adjacent to x, and $\{x\}$ is closed, it follows that $\{y\}$ is open and $y \in u_x$. Therefore $y \in f^{-1}\{f(x)\}$ and $f(y) = f(x)$. A similar argument for open $\{x\}$ and closed $\{f(x)\}$ also yields the contradiction that $f(x) = f(y)$. Our proof is now complete since (Theorem 4) every point is either open or closed. ∎

Our last result is a converse to Theorems 5 and 6.

Theorem 7. *A mapping $f: T_1 \to T_2$ from tree T_1 to tree T_2 is continuous if it satisfies (1), and either (2a) or (2b):*
  (1) *$f(x), f(y) \in T_2$ are either identical or adjacent for adjacent x, y $\in T_1$.*
  (2a) *$\{f(x)\}$ is closed in T_2 for each closed $\{x\} \in T_1$.*
  (2b) *$\{f(x)\}$ is open in T_2 for each open $\{ \in T_1$.*

Proof. Assume that f satisfies (1) and (2a). The proof for (1) and (2b) is similar. Let u be any open set in T_2. We will show that $x \in f^{-1}(u)$ implies $u_x \subset f^{-1}(u)$. This means that $f^{-1}(u)$, being a union of open sets, is open in T_1. If $\{x\}$ is open, then $u_x = \{x\} \in f^{-1}(u)$. If $\{x\}$ is closed, then $u_x = \{y \in T_1 : y = x$ or x, y are adjacent$\}$ follows from Theorem 2. We also know, from (2a), that $\{f(x)\}$ is closed in T_2. Therefore $u_{f(x)} = \{z \in T_2 : z = f(x)$, or z and $f(x)$ are adjacent$\}$. This ensures, by (1), that $f(u_x) \subset u_{f(x)}$. Since $x \in f^{-1}(u)$ implies that $f(x) \in u$, we also have (by definition) $u_{f(x)} \subset u$. Therefore $f(u_x) \subset u$ and $u_x \subset f^{-1}(u)$. ∎

Mathematical Interpretation of Beware

Each set of markers $\{1, 2, \ldots, n\}$ topologized by $\mathcal{T}'_n$ forms a tree. The playing board $Y (= A$ or $B)$ topologized as in Example 4(c) of II is also a tree. Theorem 4 shows that each point in I'_n and in Y is either open or closed. In particular, odd numbers are closed in I'_n, and shaded circles are closed in Y; even numbers are open in I'_n, and unshaded circles are open in Y. The rules of play in section I ensure that the trapper's play $t : I'_n \to Y$ and the fox's play $f : I'_n \to Y$ are continuous functions from one tree to another. To be sure, rule R_1 asserts that odd numbers are mapped into shaded circles—that is, closed points map to closed points. Similarly, rule R_2 asserts that open points map to open points. And, rule R_3 forces adjacent points in I'_n to be mapped into adjacent points in Y. Therefore (Theorem 7), both t and f are continuous. (Rule R_4 has no effect on the continuity of play; its only purpose is to force players to follow an orderly method of playing markers.) Since continuity is a basic hypothesis in fixed-point theorems, and in the particular theorems proven in this paper, any variation in the play of either player must preserve continuity. The variation of the rules in section II takes into consideration that adjacent points in I'_n may be mapped into the same point in Y. Therefore, this variation in play agrees with Theorem 6.

Is there a best strategy for playing Beware? And does there exist a marker $k \in \{1, 2, \ldots, n\}$ such that $t(k) = f(k)$? Such a marker exists for board A if the trapper covers each circle at least once (i.e., $t : I'_n \to A$ is onto). To be sure, A is a universal image (Example 4(c), Theorem 1), and so t is a universal function. Therefore t and f must agree on A for some $k \in I'_n$. Thus, the fox loses! There may or may not be a winning play for the fox when board B is used. This is clear since B is not homeomorphic to I_8 or I'_8 (Example 4(c)), and so B is not a universal image (Theorem 1). Therefore, t may or may not be a universal function. An illustration, where the fox wins, is given in Figure 7.

For the more imaginative player, more complicated boards may be chosen by simply choosing more complex examples of trees. If the playing board Z that is selected begins with an unshaded circle (see Figure 6, for

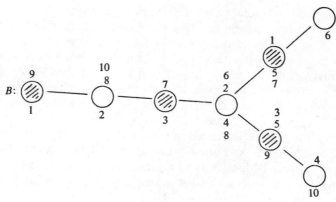

Figure 7.

example), the set of markers should be given the $\mathfrak{T}_n$ topology. Then $t : I_n \rightarrow Z$ and $f : I_n \rightarrow Z$ are continuous, and everything now hinges on whether or not Z is a universal image.

IV. Beware and Beyond

Because of its mathematical motivation, Beware can prompt questions which provide opportunities for open-ended discovery-oriented discussions. Some of these discussions can focus on the following points:

(i) Define a *monotone function* between trees to be a continuous function f for which x adjacent to y implies that $f(x) \neq f(y)$. If we consider only monotone maps between trees, does Theorem 1 still characterize universal images? Does Figure 8 below describe a universal image in this sense?

Figure 8.

(ii) A point x is an *end point* of a path P(i.e., the image of a homeomorphism $h : I_n \rightarrow P$, or $h' : I'_n \rightarrow P$, for some $n \geqslant 2$) if x is either the first or last point of P (i.e., $x = h(1)$ or $x = h(n)$). Mathematically, an end point $x \in P$ has the unique property that $P - \{x\}$ is connected. A *maximal point* in a tree T is a path $M \subset T$ that is not properly contained in any other path $P \subset T$. A point $x \in T$ is called a *junction point* if there exists at least three distinct maximal paths in T, each of which contains x as an end point. Finally, a juncture point x is said to be *proper* if each maximal path containing x as an end point has length greater than 2. In the process of playing Beware, we constructed examples of continuous, onto functions between trees with different junction point structure. A little reflection will show that it is possible to construct a universal function between two trees where the image has a junction point, but the domain does not. It is known that a universal image cannot be mapped by a universal function onto a tree with a proper junction point. It is not known whether the number and structure of proper junction points have to be preserved by a universal function.

Readers who may be interested in some background material concerning universal image spaces and universal functions should refer to [1]–[3]. The concept of universal function (and related topics) is relatively new to the mathematical community and does not appear to be in any book yet. For a reference on fixed-point theory, see [4].

REFERENCES

1. W. Holsztyński and S. Kwapien, Universal images for universal mappings, Rend. Mat. 2, Vol. 3, Series VI (1970) 197–201.
2. W. Holsztyński and F. Pedersen, Finite T_0-spaces and universal mappings, Fund. Math. XCVIII (1978) 127–139.
3. W. Holsztynski, Ordered spaces and universal images revisited, Bull. Acad. Polon. Sci. Ser. Sci. Math. Astronom. Phys. 25, no. 11 (1977) 1119–1123.
4. D. R. Smart, Fixed-Point Theory, Cambridge Univ. Press, New York, 1974.